CANCER SELECTION

The New Theory of Evolution

CANCER SELECTION

The New Theory of Evolution

James Graham

Aculeus Press Inc.   Lexington 1992

Permissions appear on page xii.

To the memory of my son.

Gregory James Graham

1958-1983

Preface

I will not repeat here the names of those individuals who helped in different ways prior to the 1983 publication of my theory in the *Journal of Theoretical Biology*. I must acknowledge, however, of my debt to the two scientists who decided that my idea was worthy of publication in that prestigious journal. Unfortunately, one of them is now deceased and I don't know the identity of the other.

The late James F. Danielli, who was then editor of *JTB*, and I had only one brief telephone conversation. An anonymous referee for his journal had rejected my manuscript and I called Doctor Danielli and to see if it were possible to obtain a second opinion. Graciously and readily agreeing to my somewhat cheeky request, he asked a more senior scientist, a member of *JTB*'s Editorial Board, to look at my paper.

Having already been rejected by the other few journals that seemed appropriate mediums for publication, I would have had no recourse if Doctor Danielli had done what most other editors would have done under the circumstances: turned me down, cold. Mine, after all, is a radical idea and he knew (I so informed him in my submittal letter) that I was not a scientist. If Doctor Danielli had refused to give me another chance, it is quite possible that this book would never have been written, and for that I am deeply grateful.

The member of the Editorial Board selected by Doctor Danielli recommended publication of my theory, but as a brief "Letter to the Editor," rather than as the full-length article I had originally proposed. The same referee later approved a second Letter that included important material I had cut from the first because of space limitations. (Both Letters are reproduced in Appendix II.)

At the time I hoped the reviewer would follow his private recommendation with an open declaration of support for my theory. He did not, however, and I can only conclude that he (I don't believe it was a woman) was not completely convinced. Perhaps he will change his mind after reading this book. But even if he is never entirely persuaded, he showed no little courage in approving an unconventional proposal developed by a rank outsider. In doing that he added greatly to my confidence.

Although I still don't know his name, there were at that time 25 scientists on the *JTB*'s Editorial Board, so for the small price of listing the names, and the 1983 affiliations, of two dozen individuals who had absolutely nothing to do with publication of my theory I hereby thank *one* of the following: S. Amari, University of Tokyo; J. Bonner, California Institute of Technology; M. Calvin, University of California; A. Cornish-Bowden, University of Birmingham; D. Garfinkel, University of Pennsylvania; N. Goel, State University of New York at Binghamton; B. Goodwin, University of Sussex; S. Kauffman, University of Pennsylvania; Y. Kobatake, Hokkaido University; R. Lefever, University of Brussels; W.R. Lieb, University of London; G. Loew, Rockefeller University; R.B. Luftig, University of South Carolina Medical School; H.M. Martinez, University of California; R.M. May, Princeton University; J. Maynard Smith, University of Sussex; J.D. Murray, University of Oxford; T. Ohta, National Institute of Genetics, Japan; H.H. Pattee, State University of New York at Binghamton; R.Rosen, Dalhousie University; M. Slatkin, University of Washington; T. Takenaka, Yokohama City University; M. Valentinuzzi, Institute of Cybernetics, Argentina; M. Volkenstein, USSR Academy of Sciences, or J.Z. Young, The Wellcome Institute for the History of Medicine.

At the early stages of my arduous but eventually successful struggle to see my theory published in a scientific journal I was quite awed by the prospect of petitioning scientists to take me seriously. There I was, with no training or experience in science and holding a mere bachelor's degree in business administration, injecting myself into an extremely broad and intensely technical field to which these extravagantly educated individuals had devoted their professional lives. I feared embarrassment, especially by making some ghastly error because I was unaware of important facts known to all biologists--or to all cancer scientists.

In an effort to elicit some tolerance for any *faux pas* I might have included in the various drafts of the theory I was mailing to a number of them, I made it clear to any biologist I contacted that I had not been trained as a scientist. This worked. Although some did not respond to my letters and some of those who did were brusque, none of the scientists who communicated with me over the years were overtly discourteous. None, that is, except for two biologists. These individuals, both of them senior faculty members at large American universities, were downright malicious.

Apparently neither of them learned what most of us are taught at our Daddy's knee: it is neither nice nor wise to inflict pain unnecessarily, particularly on strangers. Although it is probably too late to work any lasting changes in their ethical foundations or even to compensate for their failure to learn basic manners, it might be possible for these two to comprehend, that in this case, their rudeness was a mistake.

The first culprit was the referee for a biology journal, one not widely distributed. I had originally sent my theory to this journal in 1979. In rejecting it, the editor sent me the brief comments (they were typed on half a page) of an anonymous referee who, while not at all enthusiastic about my theory, was at least respectful of my effort. He even made some helpful minor suggestions. He gently told me, for example, that the correct term was "metamorphosed," rather than my more inventive solecism, "metamorphosized."

About three years later (after waiting for an intolerably long time while the editors at another journal decided, finally, not to publish it), I sent a revised manuscript back to the original journal, calling to the editor's attention a recent discovery, which had received worldwide acclaim, that strongly supported my idea. The editor accepted the revised draft for consideration but in due course

he again sent me a rejection letter. This time he enclosed four
pages of single-spaced comments from an anonymous referee, but
not from the same individual who had commented on the earlier
draft. When I saw the length of those remarks I thought that I
had at last received what I had been hoping to get from the outset:
a thoughtful analysis of my theory by a professional biologist.

How wrong I was! What I had received was a work of sadism.
His comments were intended neither to evaluate nor to criticize,
but to insult and to belittle. Reading his long sarcastic review--he
clearly spent *hours* on it--was like having my face slapped repeatedly. This professor took cruel pleasure in ridiculing my lack of
biological knowledge, my occasional lapses in English usage and
even the typographical errors in my manuscript.

There was one minor compensation. He managed, quite unin-
tentionally, to strengthen my convictions. He claimed that the
transparent arrowworms (animals I didn't know existed) were in
conflict with the theory (they actually confirm it) and he asked, sar-
castically, why I had ignored the compound eyes of earthworms
since they so strongly support my idea (which they do.) But the
useful news about the worms notwithstanding (I include those facts
in Chapters Eight and Nine), this tormentor succeeded in what he
set out to do. He made me feel like an idiot.

Typical of all bullies, he had behaved cowardly. He mounted
his vicious attack from behind the shield of anonymity that journals
routinely afford scientists who referee manuscripts. The professor
assumed he had nothing to worry about, that since I had no way
of identifying him, his mischief was without risk to him.

He made the wrong assumption.

An outsider like me, who to this day has not even met a professional evolutionary biologist, is not supposed to determine the
identity of the referee who rejected his work, especially if that
referee is a rather obscure academic. But some of us occasionally
surprise and confound. So I invite that buffoon to read this Pre-
face with care, for it contains a word which is an exact anagram
of his own surname.

The second clown is not at all obscure and he made no attempt to hide his identity from me. But he also acted in a cowardly fashion. He slipped his little piece of nastiness into a wide-ranging lecture (which he had reason to believe I might attend) but he did not mention my name and he couched his remarks craftily; I would have looked foolish had I confronted him. What he did was obliquely and briefly allude to an assertion included in my work (which had recently been published) in a highly dismissive manner. Then this well-known professor uttered a crude and ugly personal term, directed, so I am convinced, at me.

Unfortunately for him, this man, who seems addicted to fervent self-promotion, doesn't understand that hyperactivity not only cannot compensate for innate mediocrity, it actually magnifies it. He fails also to understand how imprudent it is to attack his intellectual betters.

Because I find this character, even from a distance, psychologically transparent, I am convinced that he will not be embarrassed when shown--as he is in the pages of this book--worthy of the label "Scientific Fool." But perhaps my comments on his work will encourage other evolutionists to establish standards for reasoning in the field that are equal to those that prevail, to cite just three examples, in computer programming, double-entry bookkeeping and crossword puzzle construction. If they manage that, they and their successors will have no choice but to treat the drivel published by this glib charlatan with the contempt it deserves.

Returning to more agreeable matters, Doctor John Harshbarger of the Registry of Tumors in Lower Animals at the Smithsonian Institution has shared generously, as he has since I first contacted him in 1978, material published by his office in a critically important field.

Doctor Fritz Anders of Genetisches Institut der Justus-Liebig-Universitat in Giessen, Germany sent me copies of his articles containing confirmatory findings that I have included in Chapter Eleven. I was not aware of Doctor Anders' (and his colleagues') important work and I thank him for sending me those papers.

Publishers and other owners of copyrights are thanked for permitting quotation from the following works: *Scientific American*, for James W. Valentine's "The Evolution of Multicellular Plants and Animals" (which appeared in the September 1978 issue), quoted in Chapter Eight; Harvard University Press for the material quoted from Ernst Mayr's *The Growth of Biological Thought* (1982) in Chapters Eight and Twelve; St. Martin's Press, Inc., for permission to reprint a paragraph from "Immunological Aberrations: The AIDS Defect" by Robert A. Good, from *The AIDS Epidemic*, Kevin M. Cahill, editor (an additional acknowledgement appears in Chapter Eleven); W.H. Freeman & Co. Publishers, for Errol C. Friedberg's material in *Cancer Biology* (1985), which he edited, and John Cairns' *Cancer: Science and Society* (1978), both appearing in Chapter Eleven; Plenum Publishing Corp. and Dr. Fritz Anders for the table appearing in Chapter Eleven; University of Chicago Press, for the quotations from Thomas S. Kuhn's *The Structure of Scientific Revolutions* (1970) appearing in Chapter Twelve; Dr. Michael T. Ghiselin for the material quoted from *The Triumph of the Darwinian Method* (1984) in Chapter Twelve; Alfred A. Knopf, Inc., for the material quoted from *Gods, Graves and Scholars* by C.W. Ceram in Chapter Twelve (an additional acknowledgement appears in that chapter); Dr. Michael Voselensky, for the quote from his *Nomenklatura: The Soviet Ruling Class* (1984) in Chapter Sixteen; and Academic Press Inc. (London) Ltd., for permission to reprint my two Letters published by the *Journal of Theoretical Biology* in the April 21, 1983 (Volume 101, Number 4) and March 21, 1984 (Volume 107, Number 2) issues.

S. Joyce Moore of Virginia Graphic Arts prepared the diagrams that appear in Chapters Two and Twelve.

Other than that, writing this book has been a solitary effort. I made several attempts to enlist a scientist to review the manuscript but none of those I contacted agreed to help. A letter I sent to the head of a university biology department asking for assistance with fact-checking (which I was willing to pay for) went unanswered. So be it. Any errors of fact that have crept into the book might annoy those knowledgeable readers who spot them, but they cannot be material. The most important of all evolutionary facts is my own existence and that of all other complex animals, and understanding how our complexity evolved is a matter of clear thinking, not of prodigious fact-gathering. So I ask the forbearance of specialists for any minor mistakes they may find, but I assure general readers, to whom the book is directed, that I have avoided those far more devastating errors that seem to afflict many of the fact-athletes that biology attracts. I have thought things through correctly.

I wrote this book to convince all who read it that cancer played a major role in evolution and in doing that I say nothing negative about the disease. At times I may even seem enthusiastic about its function or, at the very least, its results. This does not mean that I think it is doing anything "good" when it causes suffering and death. Although I am convinced that cancer is a biological function and that we would not exist without it, that is no reason to decrease efforts to cure anyone afflicted with it or to eliminate it. In fact, although I did not write the book as a guide to cancer researchers, it is conceivable (although I think it extremely remote) that someone in that most demanding field may find in its pages the inspiration to undertake a new approach that will prove fruitful. Nothing would please me more.

The question of whether evolution took place--the argument between Darwinists and creationists--was settled in the last century. The Darwinists won. In the following pages I frequently use "evolution-as-fact" in developing certain arguments in favor of my new theory of *how* evolution happened, but I offer no arguments to convince the reader that evolution occurred. I assume that he or she is already convinced that all multicelled organisms--those now living and those we know only from the fossils--descended from other living organisms. Anyone looking for arguments in favor of evolution versus creationism will not find them in these pages.

James Graham

CONTENTS

Part I
The Theory

Part II
Evaluations

CANCER SELECTION

The New Theory of Evolution

Part I

The Theory

One

Biology's Dirty Little Secret

All great truths begin as blasphemies.

George Bernard Shaw

Erroneous assumptions, which we may not even realize we hold, can distort our view of the world. It is only when they are verbalized or set down on paper that it is possible to modify or abandon them.

Christopher Wills

Something is terribly wrong with the theory of evolution.

The theory, one of the most pervasive and influential ideas ever developed by the human mind, claims to do nothing less than identify the mechanisms responsible for the creation of the most complicated things known to exist anywhere in the universe. It professes to understand how nature produced human beings and the other precisely constructed, highly organized and dazzlingly complex animals.

From the beginning, biology's thinkers have made *animal* evolution the primary target of their collective intellectual effort. Charles Darwin and Alfred Wallace, the two Victorian naturalists who proposed the idea of evolution by means of natural selection in 1858, are revered not because they added to our understanding of how

sea weed originated but for telling us much about the origin of
human beings and the other mammals, as well as the birds, reptiles, amphibians, fish, mollusks, insects, assorted worms and other
complex, highly mobile creatures called *animals*.

The evolutionists who came after Darwin and Wallace seemed
to add to the theory's power to explain the animals' existence.
Although Gregor Mendel, the father of genetics, worked out the
fundamental laws of inheritance experimenting with peas, the twentieth century founders of the powerful science of population genetics correctly assumed his laws applied to *Drosophila melanogaster-*
*-*fruit flies--and by extension to human beings and the other animals. Other fundamental research into the workings of heredity,
including Watson and Crick's discovery of the molecular structure
of DNA, all were assumed, without question, to add to our comprehension of animal evolution.

Contemporary theorists continue the tradition. Pick up any col-
lege textbook, visit any museum exhibit about evolution, peruse any
of the scientific journals devoted to evolutionary biology, examine
any book about evolution (those for professionals or for the general public) and you will find that animal evolution dominates.

Animals attract biology's theorists for obvious reasons. The
bilaterally symmetrical animals have brains and other complicated
body organs. They can move, many of them with spectacular profi-
ciency. Their complexity intrigues us. And, of course, we humans
are ourselves animals; what could enchant us more than the story
of our own origin?

As for plants and plant-like cell colonies such as jellyfish and
sponges,* biology's theorists seldom bother about them. They

---

*In this book I will treat jellyfish, their relatives such as coral and Portuguese
Men-Of-War (known collectively as <u>Cnidaria</u> or <u>Coelenterata</u>), and the sponges
(<u>Porifera</u>), as if they are not animals. Aristotle included these simple organisms
in the plant kingdom but biologists later decided to classify them as animals
because they move primitively. Except for that limited mobility, however, these
cell colonies have little in common with the far more complex bilaterally symmetrical animals. Assembling them with the plants under the rubric of "nonanimals" enables me to avoid an obscure technical term ("the Bilateria") or awkward
constructions such as "all-animals-except-cnidarians-and-porifera."

assume, because they are convinced that their theory of evolution by natural selection satisfyingly explains all the animals, that those other, far simpler, multicells present no additional theoretical challenge. Although I am not aware that any scientist ever felt compelled to defend their theory's claim to explain both the evolution of complex animals and evolution of the much simpler multicells, this is what one of them might say, if asked:

> What's the problem? If we've developed a convincing mechanistic explanation for the most challenging question we face--how did the complex animals come into existence--we are clearly on solid ground in claiming that the same explanation, the same set of mechanisms, are more than adequate for solving the less demanding problem of plant evolution.

That seems so reasonable, doesn't it? If a theory has the power to explain complex entities then its ability to explain simpler entities can be assumed without further amplification. It's such a modest assertion. And so logical.

But contrary to its surface plausibility, that claim is not at all logical. It's utter nonsense. It is intellectual error of the rankest sort. It is comprehensively and disastrously *wrong*. Its basic premise--explain the complex organisms and the simple ones are also accounted for--has it all backwards. Simply by restating it I can reveal just how wrong the old theory is, and how utterly unbelievable is its claim to explain, not the evolution of simple plants (which it does), but the evolution of complex animals:

> According to the theory of evolution by means of natural selection, human beings and the other bilaterally symmetrical animals, the most precisely-made and the most complex objects known to exist, were created by nothing *qualitatively* different from--nothing more powerful than--the mechanisms responsible for the evolution of the imprecisely-constructed and far simpler plants.

There it is. Biology's dirty little secret. The core theory of one of our most important sciences, the intellectual foundation for all medical and biological research, is nothing but mush. Despite the enormous accumulation of data that seemed to confirm it, because this idea claims--without explanation--that identical mechanisms caused both plant and animal evolution, it fails in its primary objective. It doesn't even try to account for the great qualitative differences between animals and plants. And any theory that cannot explain those differences does not explain animal evolution.

Just think about the plants. Unlike *all* animals, none of them have any vital organs. Their genetic material has not produced any morphological devices comparable to the wondrous organs of the animals. No hearts, lungs, digestive systems, or central nervous systems--and not a single brain cell. But the biologists want us to take their word for it; that those processes that, for hundreds of millions of years, kept plants in their primitive state also managed--Somehow!--to produce in animal lineages millions of different organisms of exquisitely precise construction, each of them replete with astonishingly complex and highly specialized organs and organ systems.

A few comparisons will show just how pretentious and downright silly this old evolutionary theory really is. According to it, the same mechanisms that created the simple mushroom also caused the evolution of monarch butterflies. Mushrooms have no organs and are utterly incapable of behavior or movement. As for the monarchs, those creatures, which weigh no more than a sheet of paper and have a brain no larger than the head of a pin, are capable of migrating thousands of miles, from summer nesting places in New England and Southern Canada, to a place they have never seen, a small site high in the Sierra Madre Oriental Mountains of Mexico. The monarch's great grandparents migrated from that specific location a year earlier!

But biology's theorists ignore that monumental difference in complexity. Those scientists insist that somehow or other the same processes that produced the simple, immovable mushrooms also caused the evolution of those prodigiously complex and precisely-made miniature flyers.

Those same thinkers also see nothing wrong in proclaiming that mechanisms which produced kelp, bean sprouts and pine trees also managed--Somehow!--to produce the whale and the elephant. Ignore--as if it were possible--that those large complex mammals have highly developed brains and an array of other precision-made complex organ systems. Pay no attention to their complex behavior, that they care for their young and interact in extraordinarily complicated ways with members of their own species and with other animals. And don't dwell on the simplicity of those cell colonies called plants. Just accept the biologists' assurances that pine tree-mechanisms explain--*completely*--the evolution of whales and elephants.

Even more dramatic comparisons can be made, of course, for professional evolutionists do not shrink from proclaiming that this "Somehow!" theory of evolution solves that most daunting of all biological challenges: the origin and evolution of human beings, the most complex of all creatures. But they seem not to be embarrassed in saying that the same processes and mechanisms that produced in other lineages nothing more remarkable than sea weed and crabgrass also managed to create the only organisms known to think. They are not discomforted by an idea that proclaims that William Shakespeare, Albert Einstein and Charles Darwin were created by mushroom mechanisms!

The old theory, called neo-Darwinism, states that the evolution of all sexually-reproducing multicellular life forms--from the simplest plants to the most complex animals--was attributable to six factors:

1. Like begot like. Organisms passed most of their characteristics to their offspring by way of the DNA transferred during reproduction.

2. All organisms had the potential to produce, and usually did produce, more offspring than those that actually survived to

breeding age.

3. Individual organisms in the same species differed from each other in ways that were inherited from their parents.

4. As a result of competing with each other and of interacting with their environments, some organisms left more offspring than others. That process, called natural selection, favored the retention of any variations that enhanced the survivability of individuals.

5. From time to time, errors, or mutations, in germ-line replication introduced innovations into the pool of genetic material. Most mutations were harmful, but some actually helped recipients to survive. Those rare favorable mutations were preserved by natural selection as part of the DNA of the lineage.

6. Because of interaction with changing environments, lineages of organisms diverged from one another. The formation of separate lines of descent contributed to the diversity of life.

All biologists agree that those elements explain the existence of all multicells including all the plants and all the complex animals. One of the six, however, is a complete fraud; it begs so many questions that it renders the theory inoperable and deserving to be characterized as a gross scientific embarrassment. I identify it later.*

---

*Colin Patterson of the British Museum (Natural History) lists only five mechanisms in his book <u>Evolution</u>, but he treats mutations and variations as one. Other biologist might compress the basic premises in other ways, but none can fault my summary of their theory's basic elements.

As for my statement that the old theory's mechanisms account for both plant and animal evolution and that biologists prefer to discuss animals, Wallace Arthur in his <u>Mechanisms of Morphological Evolution</u> confirms both of my assertions: "for no particularly good reason except my own leanings, the book is taxonomically biased in that examples are mostly drawn from animals rather than plants-...I hope...biases will not disguise <u>the generality of the principles.</u>" [Emphasis added.]

Of the six elements, the creative mechanism (so the biologists concur) was Darwin and Wallace's discovery, natural selection. As John Maynard Smith, a leading British evolutionist, wrote in 1972,

> By Darwinism is meant the idea that evolution is the result of natural selection. Neo-Darwinism adds to this a theory of heredity.

Contrary to Maynard Smith's conviction, natural selection plus heredity do not add up to a comprehensive theory. They bear the same relationship to an adequate theory of animal evolution that two slices of bread bear to a sandwich: both are necessary and important but they are far from sufficient.

If the old theory *were* sufficient, if its mechanisms comprised a comprehensive explanation of organism complexity, then some of the nonanimals ought to have become *at least* as complex as the *simplest* animals. After all, many plant and other nonanimal lineages have been around longer than the animals. They had more opportunity to benefit from the effects of the putative mechanisms of complexity-creation: they could gather a greater number of beneficial mutations, and natural selection had more time to work its creative wonders. But it simply did not happen. Despite that greater access to those (supposedly) all-powerful mechanisms *not one vital organ has ever been found in any organism, either extant or fossilized, outside the animal kingdom.**

---

*The nonanimals' sexual organs, some of which are quite complex, were obviously necessary to the survival of the lineages; however, they were not vital to the continuation of life in the individual. The old theory of evolution adequately explains complex sex organs in nonanimals. I return to this subject in Chapter Nine.

Neo-Darwinism is accepted by all biologists.* It is taught, despite the efforts of creationists in the United States, as scientific doctrine throughout the civilized world. If you don't like neo-Darwinism, but are not a creationist, you have no choice. It's the only candidate in the field.

At least that is what the professionals want you to believe. But that is not an accurate representation of the facts. There is another, completely different, *scientific* theory of evolution. It is mine. It was published in 1983 and 1984 in the *Journal of Theoretical Biology* in two brief papers that were approved for publication in that prestigious and influential journal by a scientist-referee. This new theory of evolution states, in complete disagreement with neo-Darwinism, that the evolution of humans and other complex animals was *not* caused by the same mechanisms that produced plants and the other simple multicells. It says that animal evolution benefited from an *additional* mechanism of awesome power and influence. It asserts that all animal lineages, but not those of the simple multicells, were subjected to *cancer selection*, the killing of young animals by the aggressive growth of their own cells. That assertion eliminates, with one powerful stroke, the great failure of the old theory: it explains why only the animal lineages have produced truly complex beings.

Although the professional evolutionists have completely ignored cancer,** it is the *only* biological mechanism with the specific properties needed to transform the theory of animal evolution from an unconvincing pastiche of feeble vegetable-explaining devices to a

---

*In recent years certain biologists have proposed extremely modest amendments to neo-Darwinism which they have promoted as radical challenges to it. But these singularly flaccid proposals (called punctuated equilibrium, species selection and macroevolution) have been adequately analyzed and dismissed as trivialities by more competent old-theory evolutionists. See, for example, the relevant chapters of Richard Dawkins' books, The Extended Phenotype and The Blind Watchmaker.

**Early in my theory-building effort I examined the indexes of about thirty standard evolution books. "Cancer" was not mentioned in any of them. Recently, I discovered one entry in the index of L.L. Whyte's Internal Factors in Evolution. He does not, however, assign it an evolutionary role.

plausible and coherent explanation of the origin of animals and their transformation from simple multicells to the complex creatures that now exist. I will show that without cancer selection life on Earth would not have produced anything more complex than a plant, a sponge or a jellyfish. I will convince the reader that without cancer selection, no theory of evolution can pretend to explain the existence of a single animal equipped, to use Charles Darwin's felicitous phrase, with "Organs of extreme perfection and complication."

In addition to failing to explain the complexity of animals neo-Darwinism does not account for other evolutionarily significant differences between animals and plants. These other major difficulties are also solved by my theory.

Two

Illumination With Black Boxes

Nothing can be at first sight more implausible than his theory, and yet after beginning by thinking it impossible one arrives at something like an actual belief in it.

John Stuart Mill

I will use a "black box" to explain how cancer selection caused animal evolution.

A black box is a conceptual device employed in certain types of analytical thinking. It is simply a graphic metaphor for the presumption that certain activities occur as part of a process. The theorist doesn't claim precise knowledge of the mechanisms he's placing inside the black box, he merely presumes, based on logic, or observation, or both, that they work.

To cite a famous historical example, when Charles Darwin developed his theory of evolution he had no way of knowing the mechanics of inheritance, of how the variations he observed in individual organisms were passed to offspring. Gregor Mendel, the father of genetics, *did* manage to figure it out during Darwin's lifetime, but no one paid attention to that amateur scientist's experiments; Darwin never got the news. Knowing nothing about the rules of inheritance, he was forced to develop his theory of evolution with the mechanisms for precise inheritance inside a black box.*

I use black boxes in a deliberate and literal fashion to demonstrate cancer selection's role in evolution. But before doing that I need to establish, in a black box sort of way, just what cancer is.

Cancer is a morbid process that can begin in any animal cell that normally divides. It starts when a mutational event inside the cell transforms it to a state in which it and all its descendant cells are compelled to divide in a rapid, aggressive, and destructive manner. Unless the animal musters effective natural defenses** against those malignant cells, they relentlessly proliferate in an opportunistic, vegetative fashion until they kill the organism, usually by interfering with the function of a life-supporting organ.

I've already said that mutations played a significant role in evolution. However, those mutations occur in germ line reproduction, during the transfer of genetic material from parent organisms to the fertilized egg cell which becomes their offspring. The mutational events we are now considering--*somatic mutations*--are *replication errors* that occur during *mitosis*, when a cell divides and the genetic material is transferred to the two new cells.

Unlike mutations in the germ line, which occasionally enhanced the survivability of offspring, mistakes during mitosis could *never* benefit the lineage. Even in the event--a very unlikely one--that the somatic mutation increased the survival chances of the organism, it would be impossible for that benefit to be passed to its offspring. There is an impenetrable obstacle, called the Weismann barrier (after August Weismann, the nineteenth century theorist who first suggested that it exists) that prevents changes in somatic cells in individuals from affecting cells that produce the germ line's trans-

---

*He didn't use that term. It's of twentieth century origin.

**Since I am developing a theory of how evolution worked over several hundreds of millions of years, possible medical intervention against cancer is not relevant.

mitters--the sperm and egg cells, called, collectively, *gametes*. Twentieth century research has confirmed Weismann's idea. Reproductive cells--those that will produce the gametes that pass the DNA on to the next generation--are, in animals, sequestered from ordinary somatic cells very early in the animal's life. As a result, *all somatic mutations are evolutionarily useless.**

The idea that cancer is initiated by a mutational event in a somatic cell is not new. Perhaps the first hint of that connection was uncovered in the nineteenth century when scientists discovered that X-rays could cause both mutations and cancer.** Discoveries in this century confirmed and strengthened that suspicion. One of the most significant was Bruce Ames' finding in the 1970s (which was foolishly ignored by the evolutionary biologists) that virtually *all* mutagens--the things that cause mutations--are also carcinogens. Further confirmation was obtained in the early 1980s when Robert A. Weinberg, a molecular biologist at Massachusetts Institute of Technology, identified a specific mutation that caused a human cell to become cancerous. But it was Ames' establishment of the correlation between carcinogenicity and mutagenicity that completes the constellation of factors I need to establish the black box relationships that will introduce my new theory of evolution.

My first black box represents a juvenile animal. This animal might be just a few days old and have only a few dozen body cells, or it might, if it were a preadult elephant, have already lived for a dozen years and consist of many trillions of cells. The precise age, appearance and species of the animal are not important. What *is*

---

*I am excluding from my definition of somatic mutations any mutations in somatic cells that have gametic offspring. Unlike animals, plants do not sequester cells that produce gametes from other somatic cells early in development; some germ line mutations in plants may begin as mutations in somatic cells. According to my definition somatic mutations are those passed exclusively to other somatic cells, never to gametes.

**Seven years after Wilhelm Roentgen discovered X-rays (in 1895) a worker at one of the first radiation laboratories died of cancer. He had been in the habit of testing new X-ray tubes by fluoroscoping his own hand. It was many years before the connection was made, and most of the first generation of radiologists died of cancer.

FIGURE ONE
BLACK BOX EXERCISE I

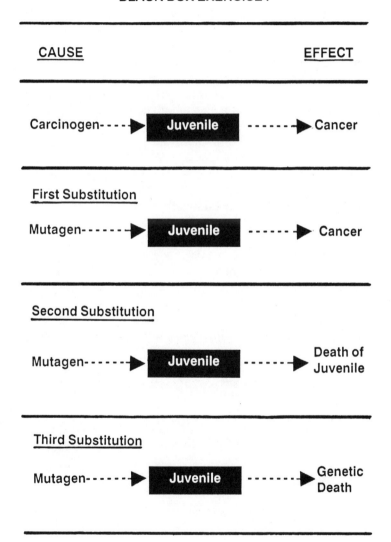

CAUSE   EFFECT

Carcinogen- - - - ▶ **Juvenile** - - - - - ▶ Cancer

First Substitution

Mutagen- - - - - - ▶ **Juvenile** - - - - - - ▶ Cancer

Second Substitution

Mutagen- - - - - - ▶ **Juvenile** - - - - - - ▶ Death of Juvenile

Third Substitution

Mutagen- - - - - - ▶ **Juvenile** - - - - - - ▶ Genetic Death important is that this imaginary animal is prereproductive, that it could not possibly as yet have any offspring.

I show the simple black box relationship between carcinogens and cancer in the illustration at the top of Figure One. If we were to place a sufficient quantity of a carcinogen into the black box (into our imaginary animal) the output would be cancer. That is a simple, straight-forward cause and effect relationship based on a self-evident truth: carcinogens cause cancer.

Since we know, from the Ames correlation, that carcinogens are mutagens, we can make an initial substitution on the input side of the black box. We can replace the word "Carcinogen" with the word "Mutagen" to describe the initiator of cancer.

Now let's move to the output side of the black box. Based on the presumption that most cancers are lethal, we can substitute, "Death of Juvenile" for "Cancer." Because the animal has not yet reproduced, any genetic material it carries for possible transmission to future generations will perish when it dies. Biologists call this extinction of genetic material "Genetic Death" and that is what, in the last substitution of this exercise, we will label the output from the black box. Our substitutions completed, what we have now is a black box showing the cause "Mutagen" and the effect "Genetic Death."

I will return to this black box later in the chapter. But at first I must set down one of the fundamental premises of my theory.

My theory asserts that there were cancer triggers or functional oncogenes in every somatic cell of every specimen of every animal species that ever existed.

The postulated presence of oncogenes in every animal cell enables us to describe a simple black box scenario of what happens inside a cell when cancer begins. There is no need to diagram this process; I will simply list the steps:

1. A mutagen-carcinogen initiates a--
2. Mutational event, which triggers an--

3. Oncogene, which causes--
4. Transformation to the cancerous state.

(Cancer researchers, who focus on actual events inside the cells of modern animals, especially humans and other mammals, will find my description of the mechanism that triggers cancer much too simple. They might point out, for example, that some human cancers do not appear until 20 or more years after exposure to a carcinogen. Or they may cite evidence that not one but several mutations must occur inside the cell before transformation takes place. I have no argument with those assertions. In fact, I explain in a subsequent chapter how my theory predicts that the cancer-initiating mechanism in any modern animal ought to be both complex and time-consuming. However, I am not attempting to describe precisely the present-day cellular mechanisms that lead to cancer. Rather, I am developing a new theory of evolution, an explanation of how animals came to exist. My highly compressed and abstract description is consistent with the evolutionarily significant cancer facts: the carcinogen-mutagen correlation, the presence of oncogenes in normal body cells [it's been established], and that it starts in one cell.)

Now for the second premise of my theory:

All animal lineages endured great losses of juvenile specimens to cancer, and most of those cancer deaths began with exposure of a single somatic cell to a mutagen-carcinogen. The theory also states that mutagen-induced lethal cancer did not occur in nonanimal lineages.

Those premises--which are supported by a wealth of physical evidence--are all I need to invite cancer into evolutionary theory. And I do not propose to sneak the dreaded killer through a side door and assign it a modest role in the history of life on Earth. No, as disturbing as some may find it, I insist on escorting cancer through the main entrance, in full view of everyone, and openly enthroning it. For I will establish beyond doubt that without cancer there would be no complex life on this planet. Not a single brain cell, no intelligence and no civilization. Without enormous numbers of cancer deaths, not even worms or insects would exist, for contrary to what we have all been taught, the other processes involved in evolution, especially including natural selection, lacked the power to cause the origin and evolution of complex animals. Cancer's brutal and efficient extermination of imperfect juveniles was central to the process.

But to begin to understand the role that cancer played in animal evolution we must do some more work with black boxes.

The next black box substitution exercise is based on the widely accepted biological principle of *selection pressure*. That well-known phenomenon can perhaps best be explained by citing the case, familiar to all biologists, of the British peppered moth, *Biston betularia*.

Until the middle of the 19th century, this moth was grey. That color gave the moths excellent camouflage protection whenever they alighted on the bark of trees, for the trees were covered with greyish lichen. Predatory birds couldn't easily see the grey moths against the grey background and tended to ignore them.

From time to time, however, the *Biston betularia* produced a black moth. The genes for black moths were rare, and because their black color made them stand out against the grey lichens, and thus visible to the birds' sharp eyes, they remained rare.

Then the Industrial Revolution dramatically changed the environment for *Biston betularia*. Factories began to belch black smoke into the English countryside. The lichens on the trees died from the pollution and the bark darkened. The moths' world had been turned upside down. Now the rare black moths were less likely to be seen and eaten, while the plentiful grey moths stood out against the black background and were easily spotted by the birds and devoured. In a few years, the population of *Biston betularia* changed. Black moths became plentiful and grey moths became rare.

The survival of the moths demonstrate the power of selection pressure. The very existence of the moths' *gene pool* was threatened by the blackening of the trees. (A gene pool is simply all the DNA in a population of related animals among whom there are no physiological or geographic barriers to breeding.) If the genes in that lineage had not produced at least a few black moths the species would have perished. But the moths' gene pool was able to survive the upheaval in the environment. It did so by making more animals with black coloration and fewer with the grey. The "mix" of genes in the pool changed in response to selection pressure.

It is important at this point to note that the surviving population of moths became black (contemporary naturalists who observered the change estimated that the population changed from 99% grey to 99% black) because *two* different events occurred over and over again. First, the black ancestors of the survivors bred. Secondly, *large numbers of grey non-ancestors were killed before they could breed.* Both kinds of events were necessary for the change to the predominance of black moths.

I have gained the impression that the second of these two *equally important* kinds of events is frequently slighted by biologists. Despite the clarity of the moths' example, the idea that deaths of nonancestors influenced evolution has a counter-intuitive feel about it. Many teaching biologists, perhaps because they do not understand it, or think it unimportant, ignore it. In typical college-level biology and genetics texts--I've looked at many of them--the authors place great emphasis on the rudiments of heredity mechanics, of how parents transfer their genes to offspring. But it is inescapable that all living things, including humans, have been determined to a large extent by the fate that befell their nonancestors. I will return to that intriguing, counter-intuitive and crucially important idea in Chapter Four, but for now I must explain cancer's fundamental role in evolution.

Because my postulated losses of juvenile animals to cancer would have had a direct effect on evolution--by killing the animals'

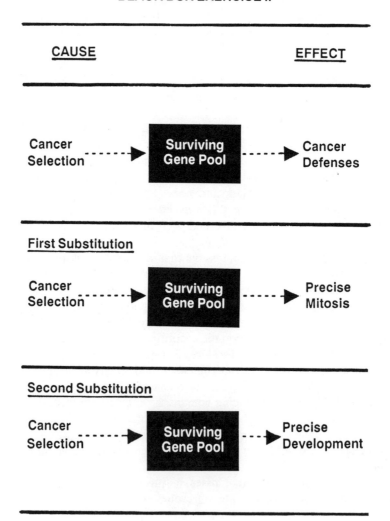

FIGURE TWO
BLACK BOX EXERCISE II

genes--I can call the event shown abstractly in Figure One "cancer selection." The cumulative effect of the pressure of cancer selection on surviving gene pools is shown in Figure Two, Black Box Exercise II.

Just as the *Biston betularia* gene pool survived because it produced fewer of the highly visible grey moths and started to produce more moths with coloration that concealed them better in the new environment, any *surviving* gene pool (any that is now creating animals) that endured losses to lethal cancer in the past responded to that threat by producing animals that did not die of cancer until they had produced the next generation.

This simple cause and effect relationship is shown in the first black box in Figure Two: when a surviving gene pool was subjected to cancer selection pressure--when cancer killed a significant number of juveniles--it responded by producing animals that were different from those that died. Just as the *Biston betularia* gene pool produced moths with the more protective darker coloring, so did gene pools under attack from cancer produce animals with better cancer defenses than those that had died. The validity of the cause-effect sequence--cancer selection causes cancer defenses--is self-evident:

Because any gene pool that did not produce in sufficient numbers organisms capable of surviving powerful and sustained threats to its existence eventually ceased to exist, we can conclude that all existing gene pools produced animals equipped with defenses against all threats that actually imperiled the gene pool in the past.

The next two black box substitutions are crucial to understanding the biological function of cancer and my theory of evolution.

Since, by definition, cancer cannot start unless something goes wrong during cell division (mitosis) it is obvious that precise error-free mitosis will avoid cancer. We can now make our next black box substitution. We can substitute "Precise Mitosis" for "Cancer

Defenses." If there was actual selection pressure from cancer in the past then it led to increased precision in mitosis.

Before making the next substitution we need to consider, briefly, the relationship of mitosis to the construction of an adult animal. Because our imaginary black box animal is a sexually reproduced multicell we can state with confidence that it began life as a single cell, as a *zygote*--a fertilized egg. We can also state that if it lived to reach the adult form it would consist of a number of cells. How many? If it were a nematode, a microscopic worm, less than one thousand; if it were an elephant, many trillions. But regardless of the number of cells in the mature animal we know that the transformation of a zygote to an adult--the process biologists call *development*--is the result of cumulative mitosis.

I have more to say about development in later chapters, for it is an extraordinarily complex and vitally important process, but for now all we need to know is:

Development is mitosis.

That relationship is self-evident. Whenever a single cell (zygote) becomes a multicelled animal (adult) the process responsible for the transformation is the repeated creation of new cells by division--mitosis.

We can make our final substitution and show that the result of cancer selection was "Precise Development."*

The validity of my substitutions, and my logic, can be judged by considering all the methods evolving gene pools might conceivably have devised to defend developing animals from lethal cancer.** There was a limited number of tactics or mechanisms available and

---

*Without precise development, "like" could not beget "like." The first element of neo-Darwinism I listed in Chapter One is the fraud.

**Like certain other evolutionists I will occasionally write as if gene pools were conscious entities. This is strictly a metaphor used only to facilitate explanation. They never were conscious.

I believe this is a summary of them all:

1. Gene pools could have protected themselves by shielding the animals from exposure to mutagen-carcinogens. By protecting dividing cells from natural mutagens replication errors (and the initiation of cancer) would have been decreased; lower error rates in cell production enhanced precise development by the genes.

2. The lineage could produce animals that would maintain extreme mitotic efficacy even in the presence of mutagens. Among modern animals, that defense is most apparent in the insects.

3. The genes could create animals using a minimum number of somatic cells in each organism. By using fewer cells per organism they would have decreased mitosis, decreased errors in development and decreased cancer risk.

4. Animals could be constructed using a significant number of somatic cells that do not divide once they themselves are manufactured. Cells that avoid mitosis cannot, by definition, be transformed into cells that divide excessively. In humans and other vertebrates all muscle and nerve cells are post-mitotic. So are almost all insect cells. Once the post-mitotic cells are formed the possibility of their having cancerous offspring is completely eliminated.*

5. Gene pools could create animals with body plans that were relatively easy to replicate. Simplification would reduce the possibility of replication errors.

6. Mechanisms inside the cell could repair damage caused by

---

*There is of course a cancer risk when the cell is formed in infancy. Some human babies die of cancer that starts in neurons.

mutagens before the oncogenes activated cancer. (Modern researchers have discovered enzymes that repair damaged DNA.)

All six of those mechanisms would *avoid* the initiation of cancer by avoiding irreparable errors during mitosis. Errors in mitosis interfere with development. Avoiding them automatically enhances precise development.

The following devices would have worked *after* initiation of steps that could lead to cancer:

7. The body could expel cells that were cancerous or potentially cancerous. In humans and other vertebrates cells in which mutations have occurred are routinely sloughed off in certain tissues, such as the lining of intestines, and leave no mutated--potentially cancerous--offspring. Some worms seem capable of *autectomy*, or self-surgery, and may use that device to rid themselves of tumors.

8. The body could acquire complex systems that would actively seek out and destroy cancer cells after their formation. That is how the vertebrates' immune systems defeat cancer.

The successful functioning of those post-transformational anticancer defenses (or any others that exist) in a juvenile would have had the same result as any that *avoided* cancer: the animal would not die of the disease; the effect of the mitosis error on its development would have been overcome and, barring other mishaps, the animal would mature and reproduce.

From an evolutionary standpoint:

The cumulative result of cancer selection in any gene pool was to enhance the DNA's control over the cell-by-cell development of animals.

Once functioning oncogenes were present in all cells and once cancer selection was established, the potential for the evolution of ultra-precise replication in developing animals was assured. Since young animals that avoided cancer death could survive and breed, those properties that enabled them to avoid, or to correct, development errors were preserved and passed on to the next generation. But if its defenses failed and the young animal perished, its genes were purged from the gene pool. Cancer selection, evolution's watchdog, ensured that *only* genetic material capable of precisely executing the development program survived and multiplied. It created a win-win condition in favor of precise development.

We can now see that cancer is a biological function:

Cancer enforced an imperative of precision. It demanded exact implementation of the genetic program inside the nucleus of all somatic cells.

My theory introduces the concept of *quality control*--the assurance that genetic material was precisely transmitted to all cells--into the evolutionary process. The old theory's emphatic assertion --that the most awesomely complex things known to exist were built without *any* evolutionarily-effective quality control mechanisms operating at the molecular and cellular levels--is an absurdity. It is an idea so lacking in worth that it warrants no further consideration by serious evolutionists. The old theory functions adequately as an explicator of the evolution of mushrooms and jellyfish, but those simple multicells mark the upper limit of its heuristic power. A theory of mushroom evolution cannot explain how complex animals came to exist.

Now I must warn skeptics; they have passed the point of no return. Unless they have found something wrong with the logic of my black box substitutions or have reason to find my postulates profoundly unrealistic their skepticism is nothing but a pose. A carelessly adopted pose. By agreeing with my substitutions they have accepted my logic. They agree with the asserted effect that cancer selection *would* have had on lineages, *if* it had occurred. Establishing that cancer selection *did* occur will be easy. All intellectually honest persons who accept the logic of this chapter and who follow my arguments in favor of the actual occurrence of cancer selection will reject neo-Darwinism.

Anyone who thinks I've made an error in my reasoning is encouraged to let me, and everyone else, know about it. I don't expect to hear from anyone, however, for rejection of my logic requires the supposition of a physiological defense against a lethal disease that begins with *imperfect* replication which would not have aided *perfect* replication. Such a defense is utterly implausible. It cannot exist. As for the possibility that my postulates are unrealistic, although I have gathered (and present in subsequent chapters) a great deal of evidence in their support, the following brief discussion ought to allay the fears of anyone who thinks I have gone too far--and dash the hopes of those who wish that I have.

Is there any evidence that oncogenes exist in all cells of animals?

Oncogenes have been found in the normal, healthy somatic cells of all vertebrate species (which include mammals, birds, reptiles, amphibians and fish) investigated to date as well as in insects and nematodes. (I will cover this subject, and other modern cancer evidence, in Chapter Eleven.)

Skeptics especially should note that although my theory was not published until after the finding of oncogenes in normal cells, it was in writing, had been sent to (and rejected by) a number of scientific journals and was registered at the U.S. Copyright Office *prior* to their discovery. I concluded *in 1978*--entirely as a result of theorizing--that animal evolution could not have occurred unless functioning oncogenes were in all cells. That's exactly where molecular biologists found them--*in 1981*.

To give some idea of the surprise and puzzlement the discovery of cellular oncogenes caused in certain quarters, this is what the British weekly *The Economist* had to say in September 1981:

what on earth are [oncogenes] for? Nature would not have evolved genes specifically intended to produce cancers. There would be no advantage whatever in that. Yet it looks as if [oncogenes] have an old evolutionary origin and have survived natural selection to climb right up the evolutionary tree of species.

The Economist's science journalist correctly saw that cellular oncogenes made no sense whatsoever in terms of the neo-Darwinian theory of evolution. But professional evolutionary biologists, who are collectively responsible for seeing that one of science's most valued theories keeps pace with all new and relevant scientific research, ignored the momentous discovery. They failed to realize that genes of great evolutionary age that routinely kill modern juveniles flatly contradict neo-Darwinism.

But if nature only retained genes that help organisms why were genes that cause cancer selected in the first place?

The genes that now cause cancer were originally selected because they performed another function, one beneficial to the organisms. In the earliest multicells, which were much simpler than animals, rapid, vegetative-like growth was not only not lethal but actually helped organisms to survive. Such growth is observable in modern plants, which characteristically grow aggressively and opportunistically, not unlike cancer cells.

Additionally, there is considerable evidence that animal oncogenes still function beneficially--by encouraging rapid growth--in the earliest stages of embryogenesis and in regenerating tissue damaged by trauma. When those routine high-growth periods come to an end oncogenes are normally deactivated by other genes.

Although my theory is about the evolutionary effects of lethal cancer and not its origin, I provide in Appendix I an expanded plausible origin scenario.

Other than the Ames correlation, is there other evidence that

mutations are involved in cancer initiation?

There is a great deal of medical evidence linking exposure to radiation and other known mutagens to specific cases of cancer. Moreover, the molecular biologists have confirmed it. Natalie Angier, a science journalist who has written extensively about the molecular biologists' research on oncogenes, has reported, "Only when the [oncogenes] are mutated do they become agents of death." I review other evidence for the mutagen-initiation of the disease in Chapter Eleven.

But if molecular biologists have identified mutations of specific genes as initiators of cancer, why would cancer selection have encouraged retention of mechanisms that reduced all somatic mutations?

If cancer is hundreds of millions of years old then nature has been moderating, for all that time, the molecular mechanisms that initiate it. Logic tells us that as animals accumulated molecular defenses against cancer the initiation process became both more complex and more time-consuming. A corollary to that conclusion is that the initiation steps were simpler and quicker-acting in the past. Furthermore, although the theory does not depend on the validity of this idea, it is entirely possible that other somatic mutations (which may have no dire effects in modern animals) initiated cancer in the distant past.

Even if initiation could not start unless the oncogenes were themselves mutated, no intelligent being looked over the shoulder of the evolving gene pools, no mentor advised them to avoid only those replication errors that caused cancer. Selection was blind. Mechanisms that reduced *all* copying errors (opaque external coverings, for example, which shield the entire animal from natural radiation) would have lowered the incidence of errors that caused cancer. They would have been selected.

Is cancer found in animals other than man?

Yes. The Smithsonian Institution's Registry of Tumors in Lower Animals gathers reports from around the world of cancer

findings in animals. The Smithsonian experts have examined specimens and other physical evidence and have determined that
cancer has occurred in mammals, reptiles, birds, fish, insects,
mollusks, and in flat worms *(Platyhelminths)*.

The finding of cancer in *Platyhelminths* is highly significant.
The most primitive of the living animals, they are possibly not very
different from the primordial ancestors of all animals. The discovery of cancer in them (which, incidentally, also occurred *after* I
had developed my theory) is strongly supportive of my claim that
the animal lineages and cancer began at the same time.

--and in nonanimals?

Not a single case of cancer has been found in any plant or
sponge. No one has initiated cancer in any nonanimal, including
the *Cnidarians*. Some marine biologists have reported finding
anomalous growths in stony corals (members of the *Cnidaria* phyla)
off the coast of Florida that may be cancerous. Since this theory
asserts that cancer selection played no significant evolutionary role
in the life history of any of the nonanimals, including stony corals,
I comment further on that ambiguous finding in Chapter Eleven.

What evidence is there that cancer starts with a single cell?

Researchers have initiated leukemia, cancer of white blood
cells, in healthy mice by the transfer of a single leukemic cell from
a mouse that already had the disease. The mice that received the
cancer cell promptly died of leukemia. In addition to that laboratory evidence, there is that frequently under-utilized scientific tool,
logic. It is much more probable that a lethal disease characterized
by rapid cell division began once in a single cell than that it began
separately in two or more cells.

Were mutagens-carcinogens present in the environment throughout the evolutionary period?

The most ubiquitous mutagen-carcinogen on our planet at this
moment is not some man-made chemical. It is not even a substance. It is ultra violet radiation. *Sunlight.* Although all extant

animals, including man, have elaborate defenses against many cancer-causing agents, including ultra violet radiation, the evidence strongly supports my presumption that sunlight was the primary cause of evolutionarily significant cancer, that it killed juveniles in astronomical numbers. I make substantive arguments in favor of heavy losses of genetic material to sunlight-induced cancer later in the book. For now, however, I will mention only four significant facts:

1. The fossils show that for about 400 million years all animals shielded all their somatic cells from exposure to direct sunlight.

2. Most animals now living never expose a dividing somatic cell to direct solar radiation.

3. The only modern animals that regularly expose dividing cells to direct sunlight (humans and certain other vertebrates) have powerful secondary defenses--immune systems--against cancer.

4. From the beginning of their life histories until now, plants and most other nonanimal multicells, which my theory says were not subjected to cancer selection, continually exposed unprotected somatic cells to intense solar radiation.

My postulate of heavy cancer selection in animal lineages and its absence in nonanimals explains that otherwise baffling historical record. Significantly, neo-Darwinism says nothing about it.

Three

The Engine of Transformation

We have...no convincing account of evolutionary progress--of the otherwise inexplicable tendency of organisms to adopt ever more complicated solutions to the problems of remaining alive.

Peter B. Medawar

Theories are like nets: only he who casts will catch.

Novalis

Now I will classify all animal genes that ever existed into four groups based on their historical relationship with cancer.

Oncogenes, which I have already described, comprise the first group. According to my theory, oncogenes have always been inside all somatic cells of organisms created by the animal lineages.

Genes in the second classification are called *anti-oncogenes*. These are defined as genes that were initially selected because they reduced genetic losses to lethal cancer; they're genes for the cancer defenses described in the last chapter. As I demonstrated in Black Box Exercise II, all cancer defenses and therefore *all anti-oncogenes functioned also as enhancers of precise replication of the genetic program during the cell-by-cell development of the animal.* That conclusion enables me to make an important extension of the theory:

*Increases in selection pressure from cancer led to increases in the
ability to create animals of complexity. The more lethal juvenile
cancer a lineage experienced, the more anti-oncogenes--the
enablers of complexity--it accumulated.*

(The direct correlation between the intensity of selection
pressure, from whatever cause, and the degree of the response to
that pressure has long been observed in nature. A good example
is the speed of cheetahs, who have been clocked at ninety miles
per hour, and the ability of new-born African antelopes to run
minutes after they are dropped by their mothers. Many cheetahs
had to die of starvation and many clumsy newborn antelopes were
had to be killed in order for those remarkable abilities to have
emerged. There are many other obvious examples of animal
characteristics, including the extraordinary camouflage of many insects, which could not have come into existence unless selection
pressure--intense and prolonged pressure--favored their origin.)

Molecular biologists have identified functional anti-oncogenes
inside the cells of modern animals. However, my definition of anti-
oncogenes is a historical one and it differs significantly from what
molecular biologists mean when they use the term. Those researchers only consider as anti-oncogenes those genes that serve that
function *now*, inside the cells of modern animals, as determined by
experimental research. (They frequently call them tumor suppres-
sors.)

As understandable and appropriate as that approach may be
to cancer researchers, I need to look at cancer defenses differently.
I am not trying to identify existing cellular defenses against cancer.
Rather, I am constructing a theory to explain 800 million years* of
animal evolution in millions of different lineages. Because I have

---

*Most works I've consulted place the origin of animals at between 600 and 700
million years ago. James W. Valentine, who says fossils can only give us an
estimate for the origin of animal life, mentions a range of from 750 to 1,200
million years ago. The figure I use, 800 million ago, is further back than many
estimates, but if I am wrong and animal evolution began later--or sooner--so be
it. The precise date is as unimportant as it is unknowable.

a different objective, most of my anti-oncogenes would not be considered anti-oncogenic by molecular biologists. I consider genes, for example, for the construction of noncellular external coverings of animals (the shells of sea turtles, the exoskeletons of insects, etc.) to be anti-oncogenes. They were selected because (1) they prevented genetic deaths from cancer by protecting the dividing cells of the animal from exposure to sunlight and, possibly, other natural radiation and (2) many animals without them died of cancer.

All animal lineages, having endured significant genetic losses to mutagen-induced cancer in juveniles, now possess a powerful array of anti-oncogenes. The reason those germ lines produce animals of exquisite complexity lies in the abundance of anti-oncogenes--everyone of them development-enhancive--that were collected because the lineages endured losses from cancer. Because of their abundance, variety and great importance--without them animals would not exist--I comment further on specific anti-oncogenes in future chapters.

The third group of genes are those that answer a question that may have already occurred to some readers (as it did to *The Economist*). If cancer came into existence 800 million years ago, why didn't selection eliminate it, at least as a killer of *young* animals? Wouldn't the selection of increasingly efficacious anti-oncogenes--defenses against cancer--have led to the elimination of the disease in juveniles? The answer to those questions is--Well, yes, certainly. *If* the old theory were correct that's precisely what would have happened. Selection would have long ago extinguished cancer's ability to kill young animals. Unfortunately for old theory advocates however, what it predicts ought to have happened did *not* happen.

Unlike the old theory, mine is correct. It states that cancer in juveniles would have been eliminated in a lineage, only if the lineage kept replicating the same basic animal over and over again, for many millions of years--in other words, if evolution in that lineage stopped. As we shall see, there are some modern animals--evolutionary dead ends--that have extremely low cancer rates for precisely that reason. But most animal lineages did not produce the same animal throughout geologic time in a never-ending stream of generations. Changes occurred. Evolution happened. Many lineages created increasingly complex animals, and *that* is why juvenile cancer occurs hundreds of millions of years after its origin.

To explain further, imagine an animal lineage that has been producing the same organism for a very long time. Because cancer selection has been steadily eliminating inefficient genes (those incapable of performing the process of development with great precision), the efficiency of the DNA in the germ line is now extremely high; the morbid process initiated by failure to replicate efficiently--cancer--has become rare among juveniles. Now let us suppose that a series of beneficial germ line mutations are inserted into the lineage's gene pool. Imagine also that the environment changes in a way that most animals without the characters called for by those recent mutations are eliminated. The lineage would very quickly start producing animals that were significantly different from those that it had been producing. What would happen to the rate of cancer under those circumstances?

According to my theory, and to logic, the rate of cancer would *increase*, concurrent with the introduction of the new adaptive character. It is contrary to logical expectations that the new-model animals could be produced with the same level of efficiency as the old models. The genes for the introduction of new characters were adaptive--they enabled more animals to survive--but they also increased cancer rates. They were both adaptive and pro-oncogenic.

But how could these *adaptive pro-oncogenes*, genes that actually *increased* cancer death rates, have been selected? Isn't *that* contrary to logical expectation? Not at all. In order to be selected, all the new genes had to do was to increase the *net* survival rate of animals equipped with them to a level higher than that of animals without the new genes.

Let's look at a hypothetical illustration to see how that would work. Assume that OLD character in a lineage has recently been significantly improved through the introduction of a germ line mutation. The mutation creates the improved model which I call NEW character. I have summarized the hypothetical survival rates (deliberately exaggerated for explanatory purposes) of animals with NEW character compared with OLD character in the following table:

Death Rate Among Juveniles

Animals with	Caused By Cancer	Other	Total	Survivors
OLD character	1%	89%	90%	10%
NEW character	10%	35%	45%	55%

NEW character animals would out-survive OLD character animals by a factor of five-and-one-half to one (55% to 10%). In just a few generations, NEW character would be in and OLD character would be out. But despite its obvious survival benefits the inevitable selection of NEW character would bring with it a *tenfold increase* in cancer death rates.

Gene pools, like hard-nosed entrepreneurs, are interested only in the bottom line. The *net* survival benefit of NEW character ensures that it replaced OLD character. And replacement would be swift. Although I have used exaggerated numbers, actual selection worked on much smaller differentials; geneticists have determined that if a mutation provides a mere 1% survival benefit it will conquer a population in about 100 generations.

Recent findings by molecular biologists support the idea that increases in complexity were accompanied by increases in cancer death rates. That research (summarized in Chapter Eleven) suggests that the more highly transformed lineages have collected more oncogenes than less evolved ones. Selection of some genes, or families of genes, for complexity apparently involved selection of more genes for rapid growth (of a larger body or a new organ, perhaps). These new genes could have acted like the original oncogenes: active in the early stages of embryogenesis and then deactivated (by anti-oncogenes), unless a mutagenic event turned them loose with fatal consequences.

But whether or not new oncogenes were added as complexity increased is not essential to this part of the theory. What is essential is that new molecular complexities added to the informational load transferred from mother cell to daughter cells during mitosis. Increases in mitotic complexity increased the likelihood of errors and the initiation of cancer.

Cancer caused by the selection of innovations would have occurred *even if the newly selected character was itself cancer-defensive.* This seemingly paradoxical situation is actually observable in humans. Both our lymph systems and our white blood cells play important roles in fighting cancer. However, lethal lymphoma (cancer of lymph cells) and leukemia (cancer of white blood cells) are a fact of life (and of death): cancer can occur in organs that fight cancer.* So it would have been for any genes that were historically selected for cancer protection. The unavoidable increase in the replication load borne by the dividing cells would have increased the possibility of carcinogenic errors in mitosis.

Because of its crucial nature, I will now offer an additional argument in favor of the idea that increases in complexity caused increases in cancer deaths among juvenile animals.

As has been the case in the past, products of human intelligence can help us to understand natural phenomena. Others have

---

*The same peculiar relationship exists in man-made cancer treatments. Radiation can both initiate cancer and kill cancer cells. This phenomenon helped me to conclude that all cells contain cancer triggers--before the molecular biologists discovered them:

Cancer cells could either die or survive when exposed to radiation, but they could not become what they already were: cancerous. Normal cells had more choices. In addition to surviving or dying they could be transformed to the cancerous state; ergo, they had within them the capacity to become cancerous. They had cancer triggers; they had oncogenes.

pointed out that man's invention of the mechanical pump helped him to understanding how the heart works. Similarly, our understanding of the nervous system was facilitated by man's invention of electric distribution systems and wire-based communication networks. More recently, computers have helped us to understand how the brain works.

Just as those and other comparisons of body organs and organ systems to man-made, multi-part machines have proven helpful to biology, man's experience in *manufacturing* great numbers of complicated objects ought to help us to understand biological evolution. In fact an excellent, well-documented record of specific experience helps us enormously. It occurred in the aircraft industry.

During World War II American manufacturers noticed that as they mass-produced a particular airplane in the great numbers demanded by the war effort, the efficacy of the process improved dramatically. As the cumulative numbers of finished aircraft increased the total worker-hours needed to manufacture each plane decreased.

Significantly, the manufacturers also noticed that if major modifications were introduced in the design of the aircraft, efficiency dropped and the number of worker-hours needed to complete each plane went up. Then, once employees became familiar with the new version, costs again trended downward.

The manufacturers eventually realized that this phenomenon-- which is now called the *manufacturing progress function* or the *learning curve*--was constant and predictable. They learned to forecast with mathematical certainty the decreases in manufacturing costs that followed the introduction of a new model.

Engineers in other industries learned that the same concept applied to their products: the initial high cost (the direct result of low efficiency) of manufacturing new products always decreased over time. The underlying reasons for that are clear and commonsensical. New models required workers to learn new procedures. Not being familiar with the new steps, mistakes were made and efficiency dropped below previous levels. But the repetitive nature of mass manufacturing ensured that the workers would eventually master the new steps, make fewer errors and increase efficiency. Learning curves work.

Now if the learning curve phenomenon prevails in human-controlled manufacturing--as common sense tells us it would--on what grounds can we *assume* that a similar phenomenon was absent in biological evolution? Shouldn't prudent theorists presume that the aggregate of DNA in a lineage at first found it more difficult to produce newly revised versions of organisms? And that efficiency increased over time? But if a learning curve did operate during biological evolution how did it function? How did dumb, blind genes learn to manufacture, with extremely low error rates, extraordinarily complex objects?

There is only one plausible answer. Lineages could have climbed evolution's learning curve only if genes that committed errors in construction were eliminated. And the only way genetic material responsible for mistakes that occurred in the replication of molecules inside the nuclei of individual cells could be eliminated is if that imperfection caused the prompt death of the animal and all of its genes. The *only* known means for a replication error inside a cell to kill a developing animal is the process we call cancer.

I have described adaptive pro-oncogenes as genes that added to the complexity of the animal in the form of a new or improved character. Actually, many other adaptive changes would have increased cancer selection pressure. We know, for example, that modern humans and the other terrestrial vertebrates live in a sunnier, more naturally mutagenic-carcinogenic, habitat than our marine ancestors. We also know that modern humans are bigger and live longer juvenile lives than the earliest mammals. Newly selected genes that enabled those changes to be expressed would have increased cancer death rates following selection.

Further support for the idea that genes for the introduction of adaptive characters caused new waves of cancer is found in modern cancer statistics for children. I comment on them in Chapter

Eleven.

Having, I hope, destroyed all resistance to the idea that past increases in organism complexity caused increases in the intensity of cancer selection, I must take care not to oversimplify. Some anti-oncogenes enabled *future* increases in complexity. Genes that provided (for example) whole body protection against solar radiation would function for hundreds of millions of years; new mutations entering the gene pool long after the shield was in place would benefit from its protection.

My theory, in other words, does not insist that *all* changes in complexity were invariably followed by increases in cancer. Which brings me to the fourth group of animal genes: *cancer-neutral genes*. These are genes whose selection was followed by neither an increase nor a decrease in cancer rates following their selection.

To sum up, the three kinds of cancer-related animal genes involved in transformational evolution worked as follows:

Oncogenes initiated lethal cancer in juveniles which created selection pressure in favor of...

Anti-oncogenes which, because of their inherent pro-replicative nature, increased the germ lines' ability both to avoid the replication imprecisions that caused cancer and their ability to execute development processes in accordance with the genetic program. Their selection decreased cancer rates until such time as changes were introduced by the selection of...

Adaptive pro-oncogenes. These genes gave the germ lines some survival benefit (more complexity, longer pre-reproductive life, movement to a sunnier habitat, etc.) but increased lethal cancer which was initiated by...

Oncogenes. More cancer caused selection pressure for more...

Anti-oncogenes, which helped the genetic material to produce the new-model animals...

And so on. Through the ages, those three kinds of genes worked collectively as a powerful *engine of transformation.* It was this cycle of cause-effect-cause, or creation-destruction-creation, that accounted for the great increase in the animal germ lines' ability to produce organisms of wondrous complexity. This powerful transformational *ratchet* explains why vital organs are found only in animals. The *kinds* of organs the animals acquired were, of course, largely determined by *natural* selection, but as plants, jellyfish and sponges demonstrate with great clarity and force, the theory of evolution by natural selection cannot possibly account for the existence of *any* organs of "extreme perfection and complication". Something else must have been involved. Cancer selection is the missing biological entity.

Four

Causes of Primordial Death

Evolution is how gene pools come to be what they are.
Oxford Surveys in Evolutionary Biology

It's probably just as well that I have never taken a university course in population genetics. That science is highly dependent on mathematics and I was never very good at algebra or calculus.* Despite my lack of formal credentials in the field, however, I will now set down three statements that form the basis for gaining an essential insight into the genetics of transformational evolution.

Statement One. All living animals inherited *all* their genes from their direct ancestors, *all* of whom lived long enough to breed.

Statement Two. No living organism inherited any genes--*none whatsoever!*--from any organism that died before it attained sexual maturity.

So far, so good. Each of those two complementary statements is obviously correct. None of our ancestors, or those of any organism, died before they bred. And any animals that died as juveniles

---

*Neither was Charles Darwin. He needed special tutoring in mathematics while at Cambridge. Moreover, the best modern texts on evolution include no mathematics and some professional theorists (Ernst Mayr, for example) have been openly contemptuous of others' attempts to apply math to evolution. I say more about this in Chapter Twelve.

could not possibly have had offspring; their genes were never passed on to other organisms. There is no need for research or argument; the two statements are true and correct.

Unfortunately, although both of those statements are valid *they are utterly useless in getting at the problem of evolution.* We need a third statement, one which is equally correct but that seems to conflict with the first two.

> **Statement Three.** The nature of all organisms now living was determined both by their ancestors *and by their nonancestors*, the animals in the lineage whose genes were extinguished because they perished as juveniles or failed to breed for some other reason.

What's going on? How can all three statements be correct? Why aren't the first two of much use in evolution?

To begin with, Statements One and Two look at organisms as *individuals*, each having an assembly of genes that were inherited, directly and solely, from their ancestors. But evolution did not happen to individuals. It happened to *populations*. And odd as it may seem, *what is true and correct for individuals is not necessarily true and correct for* populations *of individuals.*

Those first two statements also focus on *things*, on the bits of DNA that were passed--or not passed--from one generation to the next. But things did not cause changes in populations, things did not drive evolution. *Events caused evolution.*

My third statement is more useful than the other two because it is rooted in *populational thinking.* We can understand evolution only if we fully accept its populational nature.

Populations of organisms interacted with *populations* of genes, or, to use the more evocative term, gene pools. Pools of genes underwent constant changes as they moved through time, and *all* those changes were caused by *events* in the *population* of animals the genes themselves had created.

Events in animal populations that changed gene pools are precisely classifiable. Animals--indeed any organisms--could, in the final analysis, perform one of two--and only two--evolutionarily significant acts: they could breed, or they could not breed. As far as evolution in their own lineage was concerned nothing else mattered.*

Although most conventional theorists emphasize the mechanics of direct inheritance of characteristics (possibly because it is more easily observed in laboratories and taught in classrooms), all juvenile deaths and reproductive failures that occurred in surviving gene pools, especially those that happened in the more remote past, most certainly influenced the course of evolution.**

Why do I say that events in the *deepest* past were more evolutionarily significant than more recent deaths? We can get at the answer to that question by considering our earliest known multicellular ancestors which, so the paleontologists tell us, were Pre-Cambrian worms. How many descendants did those founding fathers and founding mothers of the animal kingdom have in the hundreds of millions of years since they lived and died? We have no way to count them but they most assuredly were in the astronomical range--in numbers so huge that words like quadrillions, quintillions and even googols are inadequate. Now consider what would have happened if a single one of those earliest ancestors, a worm that uniquely possessed a then-new beneficial gene, one that has actually survived to the present time in all living animals, had

---

*Those familiar with the sophisticated concept of kin selection might think that some animals can do something else that affects evolution: they can help their close relatives to live and breed. Sterile worker bees, for example, commit suicide when they attack animals that threaten the hive. According to kin selectionists, the bees do this to ensure the survival of copies of their own genes located inside their mother, the queen bee, or in siblings who may reproduce in the future. I have no argument with kin selection but suggest for purposes of understanding this chapter that it be viewed as a special form of breeding. Perhaps breeding-by-proxy is an appropriate term. In any event, kin selection occurs only in the more complex animals and all aspects of animal complexity, including altruistic suicides in bees, owe their existence to cancer selection.

**Is it possible that some biologists do not acknowledge the importance of genetic deaths because they do not understand basic evolutionary mechanics? Anthony Smith in The Human Pedigree, in writing of premature human deaths, assures his readers "...babies (that) die...might as well not have lived for all the genetic effect they will have upon the future of mankind." More amazingly, the current President of the Society for the Study of Evolution, Stephen Jay Gould of Harvard University, in writing of mutants that do not survive (in an essay included in Perspectives on Evolution [Roger Milkman, editor]), confidently tells his readers, "Most monsters, of course, are hopeless...and these play no role in evolution." These glib remarks suggest two fundamental mistakes: the biologists thought in terms of individuals instead of populations and they did not accept the historical nature of evolution.

instead died as a juvenile. Because of its role as a founding organism--with an astronomical number of descendants--all animal life on this planet would have been drastically altered--by the death of that one worm!

Now let us pose a slightly different question: What would have happened to future animal life if a single worm that in fact *died* as a juvenile because it lacked a single gene that the survivors possessed, had, instead, lived and bred? The answer, of course, is the same: the nature of life on this planet would have been vastly different. If that single immature worm had not died in the Pre-Cambrian the evolution of animal life would have been altered beyond recognition.

As those examples show:

Evolutionarily significant events that occurred in the deep past were more momentous than more recent events because the number of descendants (actual or potential) affected by the event was immensely larger.

I find it helpful to understand the great significance of events in the deep biological past by comparing evolved gene pools to interest-bearing bank accounts. Although that analogy might startle more conventional thinkers, gene pools and interest-bearing accounts share important attributes. Because of the effect of compounding, which can convert a single dollar to an astronomical sum in a mere two thousand years (in evolutionary terms, a blink of the eye), transactions made in the earliest years of such an account's life would have had far more impact on the current balance in the account than those of more recent occurrence. If a single imaginary dollar had been deposited 2,000 years ago in a bank account on which interest compounded at the rate of 5% per annum, that dollar would now be worth $231,000,000,000,000,000,-000,000,000,000,000. And one dollar *withdrawn* from such an account 2,000 years ago and squandered on some frivolity would have inflicted upon the current owner of the bank account an actual economic loss equal to that same enormous amount.

Much to the relief of those responsible for maintaining order in our economic affairs, there aren't any 2,000 year-old interest-bearing bank accounts.* But there is nothing imaginary about the cumulative nature of extant animal gene pools, all of which are hundreds, or even thousands, of millions of years old. Evolutionary theorists must deal with reality. Significant biological events--breeding and nonbreeding by *real* animals--actually happened 800 million years ago. Those deeply distant events had profound effect on current life forms; unlike make-believe transactions in imaginary bank accounts their influence was real.

Those who have been indoctrinated with the notion that "real" science deals only with phenomena that can be observed and measured may not like it, but anyone who is serious about evolution must confront the great significance of long-ago events. Their influence must be fathomed. We must, somehow, go back in time. And it can be done. We can use legitimate intellectual devices to penetrate the fog of deep time and "observe" those remote but highly determinative evolutionary events.

The idea that we can *know* anything useful about events that occurred hundreds of millions of years ago is counter-intuitive. So is the assertion that whatever aborted the short lives of doomed primordial juveniles greatly influenced living animals. Both ideas are correct, however, and both are essential to understanding evolution. To demonstrate their validity I will conduct a thought experiment.

I will start my experiment with an animal we all know, the elephant. I will show that we can reach many sound conclusions about actual events that took place in the elephant lineage's deep past. To do this we must think clearly and supply; now with boldness, then with caution. And we need make only one very modest presumption: that only biological mechanisms we can observe today were functioning throughout the lineage's history.

If we travel back in time, in our imagination, and think about

---

*An account of that age would have long ago become a monetary black hole, sucking in more money in annual interest than any economy could possibly produce. Benjamin Franklin, who understood well the effect of compounding (and so many other things) established two interest-bearing accounts (each with initial deposits of one thousand dollars) for the future benefit of citizens of Boston and Philadelphia. He specified that the funds remain on deposit for two hundred years. By that time (it was in the 1980's) the thousand buck deposits had grown to millions. But the accounts were not yet unmanageable monsters.

what the elephants' ancestors looked like, we might start by trying to conjure up their immediate ancestors. Those were most likely also very large four-legged mammals who survived by grazing. They probably had tusks and trunks and otherwise looked a lot like modern elephants.

But to understand the more important evolutionary events we need to leave animals that were similar to elephants and go back further in the lineage. And as we travel back the animals will look less and less like elephants. Paleontologists tell us that the first mammals were small. They looked more like mice or rats than elephants. Further back the animals will be even more different for they won't even be mammals. Reptiles, amphibians and vertebrate fish would all appear eventually in our imaginary travel backward in the elephant's lineage, and in that order.

Finally, if we reach the very earliest animals, perhaps as much as 800 million years ago, we would come to the founding species of the lineage. We have no way of knowing what those original animals actually looked like--there are no fossils--but the oldest animal fossils yet discovered are those Pre-Cambrian worms. Worms that lived in burrows at the bottom of the sea. Those animals were relatively sophisticated--burrowing demands complex musculature and a central nervous system--and large; some of them were one meter in length. The first actual animal ancestors of the elephants were probably simpler and smaller than the worms, but, since we will probably never learn anything about those original animals, we can let the worms serve as their proxies. They lived a very long time ago, they were greatly different from, and far less complex than, the elephants. They will do.

Having reached that presumed starting point of the worm-to-elephant lineage we can now start to identify the determinative events that took place in the lineage.

But before explaining how I deduced those past events, we need to understand better the relationship between a lineage of evolving animals and that lineage's gene pool.

Throughout the hundreds of millions of years of animal evolution, two biological entities--gene pools and populations of organisms--travelled through time interacting with one another. Gene pools created animals and the animals caused changes in the gene pools either by breeding or not breeding.

I find it useful to think of animals influencing evolution by providing informational *feedback*. The gene pool of those British moths, *Biston betularia*, received feedback every time a grey juvenile moth was eaten by a bird. The message? *Grey coloration is not working.* Black moths who survived bird predation and bred also communicated with the gene pool. The breeding black moths delivered this message: *Black coloration is working.* The gene pool reacted to the cumulative messages from the non-breeding grey moths and the breeding black moths by doing what any "intelligent" reactive entity would do: it increased production of black moths and decreased production of grey moths. The *Biston betularia* gene pool "learned"--from events taking place in the real world inhabited by the moths it created--to produce a new "mix" of animals, one that made survival of the gene pool more likely.

I am, of course, using a metaphor. There was no "learning," not in the sense that humans or animals learn. There was no intelligence at work. Moths either lived and bred, or they suffered genetic death. The gene pool was never conscious and didn't learn anything at all in any literal sense. It did, however, change. It evolved. And just as the pump metaphor enables us to understand the heart, so the feedback-as-learning metaphor enables us to understand the relationship between gene pools and the animals they created. All extant gene pools *responded* to significant *past* changes in the animals' environments--changes that placed the gene pools' existence in jeopardy--by "learning" to produce populations of animals that were more likely to survive and breed.*

How may we use this understanding of the relationship of evolving gene pools to evolving animal populations to determine what genetic events occurred long ago in the worm-elephant lineage? Actually, the problem is not as difficult as it may seem, for those of us who work in the past have certain advantages over those who deal with the present and the future.

1. We *know* we are dealing with events that actually took

---

*This phenomenon has been directly observed in the reaction of bacteria to antibiotics and of insects to insecticides. Strains of organisms resistant to chemicals designed to kill them now exist because their gene pools reacted to those actual threats to their existence exactly as I have described. In the thirty years since DDT was introduced, for example, more than 200 lineages of insects developed organisms which are immune to it. Those resistant strains would not have evolved unless many billions of insects actually died of DDT poisoning.

place. There is a completed series of events. We do not need to predict, or speculate about, the future. Although we cannot actually see the events that transformed worm genes into elephant genes, we can, by thinking imaginatively (but carefully), benefit from the certainty that those events are over and done with.

2. We can presume that the descent of elephants from worms was a straight line process and that each species descended from only one other species and were not products of hybridization. Biologists know that animal hybrids are usually not fertile. Mules, which are half-horse and half-donkey, are the best known example of sterile hybrids.

We can never know what the animals in all the species (hundreds of thousands? millions?) in the worm-to-elephant lineage looked like, but we can presume, based on what we know about present-day animals, that the species came into existence one at a time.*

3. The presumption of lineal descent means that all the genetic changes in the worm-to-elephant lineage occurred in *complete genetic isolation* from events that took place in other lineages.

I find it helpful to think of those isolated worm-to-elephant genetic changes as having occurred inside a very long tube, one that's 800 million years long. The tube is sealed at both ends. It is closed at the front end, the beginning of the lineage, because I have no theoretical interest in what happened before the worms came to exist. And it is sealed at the other end, the one with genes producing present-day elephants, because I am not going to

---

*In the unlikely event that two different species did interbreed to create a new species, they would have had to be closely related to each other. Perhaps animals with common origins, like lions and tigers, could mate and produce fertile offspring. But neither lions nor tigers could mate with alligators. Because of the need for any two parents of a potentially fertile hybrid to be from closely-related species, the logic I use in this chapter is valid even if hybrids played a more significant role in animal evolution than is generally thought. Any lineage that may have temporarily developed two lines of descent and then merged into the hybrid species would have reestablished the now-completed lineage at the point of the last hybridization. The uniqueness of each lineage, which is essential to further development of the logic in this chapter, was a certainty.

speculate about the elephants' future. Since it is closed at both ends I can state in confidence that *all* the genetic events that occurred in the transformation of worms to elephants are inside that tube.

We might think of the worm-to-elephant gene pool and the trillions of animals it produced as having been pushed through that long and narrow tube by a time-driven piston. Throughout that very long passage, from worms to elephants, the genes were producing animals (whose characteristics, except for the beginning and ending species, are unknowable to us) and all the animals were continuously feeding back information to the gene pool by the only means possible, by breeding or by not breeding.

Now I can make certain deductions about the genetic events that took place inside that sealed tube, the events that changed worms to elephants.

All genes in the tube at the beginning were worm-genes. Over time, new genes, which originated as germ line mutations, were added. Those two sources--worm genes and mutations--contributed all the raw material available for selection during the lineage's 800 million-year lifetime.*

Many of the genes in that mass of raw material went out of existence because animals that possessed them did not reproduce. In other words, genetic deaths occurred.

I can now prepare a severely compressed, but most useful, summary of all genetic changes that occurred during worm-to-elephant transformation:

Begin: **Worm genes.**
Add: **Germ-line mutations.**
Deduct: **Genetic Deaths**
End: **Elephant genes.**

---

*Recombination of genes, a reshuffling which occurs in every act of sexual reproduction, played a crucial role in evolution. However, recombination does not create new genes or kill old ones. I can ignore it for purposes of this exercise.

Population geneticists, who develop data in experiments to which they then apply sophisticated mathematical techniques, will not like that summary. But those laboratory scientists are concerned with observable, and thus decidedly *minor*, transformations. They observe, for short time periods, small populations of organisms--usually fruit flies--kept in laboratory containers. They always start with fruit flies and they always end with fruit flies. But I start with worms and end with elephants. I cannot see minor changes occurring in very short time periods in small populations. I cannot observe and measure specific gene changes. I need a technique that differs significantly from the geneticists'. My approach *must* be conceptual, my method, inferential.

The geneticists measure gene frequencies in laboratory populations. They might, for example, measure the percentage of flies in a controlled population that carry genes for small wings. But when it comes to solving problems of large scale transformation, the differing gene frequencies at any moment in a specific population are as immaterial as they are unknowable. Certainly, gene frequencies changed in the Pre-Cambrian worms' gene pool and of course there are different gene frequencies in the elephants' gene pool. And if we could have monitored the changes in the gene pool throughout those 800 million years, the first sign that a particular gene was going out of existence would have been a reduction in its frequency in the population. But I have no access to (or theoretical interest in) past fluctuations. The technique I have used is the right one. I have telescoped time and forcibly classified all the large-scale changes in the worm-elephant gene pool into clearly differentiated categories.

And I compress and categorize with utmost confidence. The historical nature of evolution assures us that *all* the genes that ever existed in the lineage either perished before elephants came to exist or they survived and are now elephant genes. My summary has captured and correctly identified all the relevant *kinds* of genetic events that occurred in the lineage.

Of the two kinds of changes--germ-line mutations and genetic deaths--I am more interested in the latter because I can determine *causes* of evolutionarily significant genetic deaths. I can do that through my understanding of how selection works and by applying the power of deductive reasoning.

To begin, we need to state how genes died in any lineage:

> *Genes went out of existence only if something bad happened to the animals that carried them.*

Genes never survived in nature separate and apart from the body of an organism. If genes died in the worm-elephant lineage (and in 800 million years they died in astronomical numbers) then they perished because of some failure in the animals that carried them.*

There were only two types of organism-failure that killed genes: reproductive failure and organismic death.

In reproductive failure animals survived to maturity but failed to have any offspring. Others had some genetic flaw, or handicap, and left fewer offspring than competing individuals. In either case, whether it occurred in a single generation (no offspring) or over a small number of generations (fewer and fewer offspring) the end result was the same: genetic death caused by reproductive failure. Genetic deaths from that source contributed importantly to the evolution of animals and other sexually reproducing multicells.

The other failure-type--the actual deaths of juvenile animals-- is the more common of the two modes of genetic death. We know that because of the high number of animals born in every species compared to the much smaller number that survive to breeding age. In Darwin's words, "many more are born than can possibly survive." As I will explain, we can use that fact to determine causes of death of many juveniles, even those that took place in the deep past. This is possible because *animals usually die of a single cause.*

If someone were to ask me to list all the physical events that explain why I am alive today, I wouldn't know where to begin. My body has survived because many trillions of crucial events have taken place during the decades that passed since my conception. Heart valves opened or closed at the correct microsecond, sophisti-

---

*To preclude objections from those familiar with genetic drift and junk DNA: Yes, I know that some genes may float in and out of gene pools without the influence of selection. However, until it is shown that these bits of DNA have significantly influenced phenotypes by, for example, creating tissues or organs, I can ignore them as nothing more than the biological equivalent of background noise--evolutionary irrelevancies.

cated thermostatic mechanisms maintained my body temperature at an ideal level, my digestive system produced chemically complex enzymes enabling me to digest life-sustaining food, my immune system destroyed life-threatening foreign organisms, and so forth. *Ad infinitum.* The list would be endless, literally impossible to compile.

If, on the other hand, I were to die tomorrow it would probably take a competent pathologist an hour or so to figure out what had happened. Unless I had been killed by a rare pathogenic micro-organism or an untraceable poison, he or she would quickly isolate the cause. And it would probably be a single cause. Pathologists have found that even very old humans usually die from one identifiable failure; all their organs may be weak, but in the end it is the dysfunction of one organ that terminates life.

In developing an evolutionary theory our problem is different from that of the pathologists'. We cannot conduct autopsies on the quintillions of individual animals that died over hundreds of millions of years; the past remains inaccessible to direct observation. However, we can, through our understanding of selection and by populational thinking, nonetheless *know*, with certainty, the causes of many juvenile death.

We can ascertain cause of primordial death because natural selection "works" on variations among individuals in a breeding population. We also know that selection consumes--eventually eradicates--those differences.

Given enough *time and* sufficient *selection pressure, all evolutionarily significant variations among animals in a species are eliminated.*

For example, if the trees in Britain remain black and if the predatory birds remain sharp-eyed and hungry for moths, the grey-colored *Biston betularia* will *eventually* disappear. It would take a very long time but selection pressure would eliminate them. And if they did in fact disappear, a future scientist who looked at the black moths could--even if he knew absolutely nothing about *Biston betularia's* history--conclude that all non-black moths in that lineage *must* have suffered genetic death in order for the species to consist entirely of black moths. This future theorist would not know the color of the "disappeared" moths or *when* they ceased to exist.

Nor would he know *what* caused the elimination of the nonblacks, although he might, if the trees were still dark, and the birds still insectivorous, make a fairly good guess. But he could deduce, with confidence, that--*at some point in the past, in that lineage*--all nonblack moths died without offspring. That conclusion would, for all practical purposes, be as valid as any scientific truth derived from laboratory observation. (The only alternative would be that *Biston betularia* was the founding species of all moths, that it has produced nothing but black moths from its inception and that all other moth lineages, most of whom produce nonblack moths, descended from it, but didn't start to produce nonblack moths until their own separate lineages had been established. I think most readers would agree that this sole alternative is far-fetched.)*

We can reason similarly about other characters in other animals. I recently watched a bright orange-colored butterfly alight on the ground a few feet in front of me. When I walked closer to get a better look at the creature all I could see were brown leaves lying on the grass. I brushed at some of the leaves with my foot to see if the insect had crawled under one of them when, suddenly, one of the leaves flew away. The "leaf" was the butterfly!

That butterfly species developed sophisticated camouflage on the underside of its wings that fooled me (as it undoubtedly has fooled many hungry birds) into thinking it was an autumn leaf. I can conclude--I *know*--that a lot of butterflies in that lineage died because they did not possess that particular, highly effective, type of camouflage.

By judicious application of that kind of deductive reasoning we can determine, with confidence, the nature of many evolutionary events in the past. The following are some examples for the worm-elephant lineage.

All elephants have highly functional trunks. We can infer from this that all animals in ancestral species within the lineage that had no trunks--or inadequate trunks--were eliminated. This is *not* to say that trunk-less animals did not flourish in the lineage

---

*Actually, the life history of Biston betularia is not following the scenario I describe. Efforts to curtail pollution in Britain have lightened the bark again and the grey moths are making a comeback.

in the past. And it is not to say that *other* animals in *other* lineages did not descend from those trunk-less ancestors of elephants. But the survival of trunk-less animals at earlier times or in collateral lineages does not in any way negate the inference that in the elephant's lineage *all* trunk-less animals--everyone of them--eventually succumbed to genetic death.

Based on the physiology of present-day worms, it is a safe bet that the primordial worms had a simple circulatory system and, as is the case with modern worms, that it was driven by a single-chambered heart. But all elephants have four-chambered hearts. We can conclude that all genes used exclusively in the construction of single-chambered hearts died in that lineage. They died hundreds of millions of years ago and their deaths were not witnessed by any humans but that doesn't matter. They most assuredly died.

Elephants can digest large amounts of cellulose. Conclusion? The *in*ability to process cellulose was responsible for many genetic deaths in the lineage.

As those inferences suggest, we can formulate what I call the "cause of death rule":

Whenever we observe a specific complex *and* adaptive physical phenomenon *that is* universal *in all members of a group of animals we can conclude that lack of that particular feature--functioning at its current level of efficiency--*must *have been the cause of many past genetic deaths in that lineage.*

The reader should note that I have not said anything about the *mode* of genetic death.   Most proto-elephants with insufficient trunks probably died as juveniles because they were unable to obtain sufficient nourishment. But it is remotely possible that some survived to the adult stage but their less-than-adequate trunks made them unappealing to prospective mates. It doesn't matter how they died or even if some genes originally for no-trunks continue to survive in present-day elephants with trunks (where they do something else.) The analytical instrument I am developing in this chapter is powerful, but, at this stage, it is also crude.

We cannot see the elephant's nonancestors' genes dying because they produced animals with inefficient trunks. We cannot observe animals dying because their genes didn't enable them to digest cellulose. We cannot give a name to the ancestral species in which they died, or identify the geologic era in which they perished. But we can state with certainty that genes used solely to produce now-nonexistent characters died.

We can expand the "cause of death" rule to accommodate major events in the lineage's history.

Elephants descended from animals that emerged from the sea and colonized land. We can conclude that genes used only in the production of characters used for marine life died.

Elephants descended from worms. This informs us that all worm-only (or fish-only, or reptile-only) genes were exterminated by genetic death. The fact that worms, fish and reptiles now exist does not conflict with that conclusion. Inside the enclosed tube all genes used exclusively to produce nonelephants were eliminated.

We can in fact, after scrutinizing the elephants' anatomy, observing their behavior and contemplating their evolutionary history, develop a very long list of specific causes of genetic death in the lineage. We can do the same thing for any animal and our conclusions would be as valid as those based on direct observation. We *can* penetrate deep time.

The usefulness of this reasoning will become clear in the following chapter when I show that the fact of animal evolution leads inexorably to the conclusion that it was caused by cancer selection.

Five

Wondrous Unbroken Chains

It is an interesting historical fact that in this century the great
advances in evolutionary biology made possible by the population
geneticists completely ignored the role of development.

John Tyler Bonner

In Chapters Two and Three I argued from logic to fact: *if* cancer had caused a lot of juvenile deaths in the past *then it* would
have caused animal evolution. In this and the next chapter I argue
from facts to logic, that the reality of animal evolution says my
theory is correct.

To start, here are a few straightforward facts about animal
evolution:

All animals descended from other animals.

Animals have existed in great numbers. No one knows how
many lived and died in the 800 million years since they originated, but we can gain some sense of the numbers by
considering that the *present* insect population has been
estimated at a thousand million thousand million individuals.
Another indicator of magnitude is the consensus that all the
world's petroleum and much of its natural gas was formed from
the decomposition of *some* of the animals that lived in the

distant past.

But, awesomely large as their numbers undoubtedly were, all the animals that lived in the past can be placed into two distinct groups:

1. Breeders.

2. Non-Breeders.

I classify with impunity. All now-dead animals either left *some* offspring or they left *none*. This simple categorization enables me to infer certain indisputable facts of theoretical value about one of those immense groups. This is what we know about the **breeders**:

> They were all alive when they engaged in the sexual acts that produced the offspring.

> All were adults when they bred.

> Being adult *animals*, all had complex bodies complete with functioning organs.

> All those adult animals were products of complex development *processes*. They began life as zygotes and, following numerous cell divisions (many trillions in the larger specimens), became adult multicells.

> This self-manufacture of adult animals was controlled by a genetic *program*, a precise set of instructions embedded in the animals' DNA.

We can expand the list to include another indisputable fact, one of great theoretical import, about those *breeding* animals:

> Evolution from Pre-Cambrian worms to elephants (and to dino-

saurs, insects, humans, etc.) required massive revisions, over time, of both the (controlling) development *programs* and the (controlled) development *processes*. Not only was the DNA inside each zygote (what biologists call the *genotype*) subjected to enormous change over evolutionary time, but so was the *process*, the actual cell-by-cell manufacture of the animal. (The actual organism is the *phenotype*.)

I have summarized those particular evolutionary facts in that original way in order to show how truly stupendous animal evolution was:

> *Despite prodigious genotypic and phenotypic transformations, in all animal lineages throughout all of evolution the genetic instructions for construction of animals that eventually bred were implemented* without exception *with the extreme precision needed to produce a viable adult complete with organs and capable of mating with another, equally complex, conspecific.*

The reality of those wondrous unbroken chains of precise implementation of genetic programs is *not* speculation. It is *not* hypothesis. It is *fact*. And what a devastating fact! In all the greatly transformed extant animal lineages--millions!--the chains remain unbroken now for hundreds of millions of years!

The fact of the unbroken chains is not only *not* explained by the old theory, it is not even identified as a problem! But it is just as well that it isn't. Nothing neo-Darwinists can say could explain it. No Darwinian struggle for survival accounts for it. The environment certainly didn't do it. Geneticists' research cannot help. Years of carefully designed laboratory experiments with fruit flies would yield no clues as to how they came to be. And regiments of biologists' formulae could never offer even a shadow of the correct answer.

No, that great mystery can never be solved with neo-Darwinian mechanisms. My theory, on the other hand, explains it fully.

The explanation for faultless self-manufacture in all the breeding animals lies, of course, not in them but in that other large group of now-dead animals; the answer is in the **nonbreeders**. We need to think fruitfully and clearly about all those animals that died without offspring to understand what happened in those that lived to breed.

I laid the groundwork for the logic I am about to use in the last chapter. Using obvious examples (elephant trunks, camouflage in insects, etcetera) I showed that specific causes of past genetic deaths--sources of selection pressure--can be determined by thoughtful consideration of organisms that now live. In working with those examples, however, I refrained from identifying with certainty the mode of genetic death; although juvenile death was the more likely culprit, reproductive failure was a possibility. But in attacking the problem of developmental fidelity in all breeders I need not hesitate. Now I can use the power of the logic with precision.

First of all, recall what the cause of death rule enables us to do:

Whenever we observe a specific complex *and* adaptive physical phenomenon *that is* universal *in all members of a group of animals we can conclude that lack of that particular feature--functioning at its current level of efficiency--*must *have been the cause of many past genetic deaths in that lineage.*

Genetically-determined development fidelity is observable in all living animals and, to evolutionists, it is as good as observable in all the now-dead breeders. This universal, adaptive and complex phenomenon tells us that insufficient fidelity *during* development somehow caused many genetic deaths in the past.

Because developing juveniles could not breed we can eliminate reproductive failure as the mode of the genetic deaths responsible for precise development. But still-developing animals could have

altered their gene pool and profoundly influenced evolution by performing one particular act. They could have *died* during development. And the fact of those great chains of precise replication
enables us to say what killed many of them.

The fact *of replication fidelity in all breeders proves that in all
animal lineages great numbers of juveniles died of replication in-
fidelity.*

And, by the process of elimination--

Mutation-initiated cancer killed them.

To demonstrate the validity of those two statements let's
examine all other imaginable explanations for the breeders' universally successful development. These seem to be the only alternatives:

Alternative 1: During all of animal evolution no replication
errors ever occurred during development. No mistakes in breeders
and none in nonbreeders. Not only were animals in evolutionarily
stable populations invariably produced without replication failure,
but all *changes* mandated by evolutionary transformations were
faultlessly integrated into the DNA and *faultlessly* implemented
during development.

Alternative 2: Replication errors occurred during development
but the mistakes never caused death of the animal. Barring other
mishaps, all those misreplicated animals became breeders.

Alternative 3: Errors in development did occur, and juveniles
died as a result of those failures, but all the deaths were caused by
replication errors above the level of molecules inside somatic cells.
In other words, some animals died because the development process failed to construct tissues and organs in strict accord with the
genetic program, but--so this alternative would claim--every somat-

ic cell (in some individual animals, many trillions) invariably contained (a) a sufficiently precise replica of all the genetic material inherited from the animal's parents and (b) all the vital physio-morphological characters mandated by the development program for that particular cell type.

Alternative 4: Errors in cell replication during development occurred and they caused the death of many developing animals-- but the animals all died of something other than cancer.

We can dismiss the fourth alternative first. If a widely observable mechanism has all the characteristics the missing selection agent would need, no serious theorist would propose a nonexistent substitute.

Each of the remaining three alternatives presumes, implicitly or otherwise, that God did it: that an omnipotent nonbiological entity ensured that the genetic program was executed at the molecular and cellular levels with mind-boggling precision or, if mistakes were made, He fixed them.

If the Creator exists we can be confident that His omnipotence enabled Him to maintain the chains of precise replication. But for those of us who consider only natural phenomena feasible, cancer selection is, in this case, God's equal. It too is perfect. Omnipresent oncogenes would have ensured that the process of gene elimination could have begun in any cell misreplicated during development (including those that do not normally divide again) and it would have worked fast enough to kill juveniles.*

Uninterrupted precise development of complex, organ-equipped breeding organisms during prolonged lineal transformation *happened*. Because cancer selection is the only coherent natural explanation for that reality, the fact of animal evolution proves the

---

*Neo-Darwinists may not appreciate seeing their theory excluded from consideration while creationism is treated, albeit briefly, as a serious alternative to cancer selection. But a theory that doesn't even recognize the unbroken chains as a problem is asking to be treated as a befuddled bystander.

validity of my theory.

Six

Evolution of
Animal-Manufacturing Systems

Something deeply hidden had to be behind things.

Albert Einstein

Gene pools, which I earlier defined as simply all the DNA in an interbreeding population, were, in animal lineages, much more than that. Throughout the long evolutionary process they masterfully crafted those extraordinarily complicated objects called animals. Not only do present day gene pools control the construction of those amazing things, but they produce them with predictable ease and in astronomical numbers. So impressive is that accomplishment that the central question of biology ought not to be "How did complex animals evolve?" but rather, "How did manufacturing systems capable of mass-producing complex animals evolve?" No theory of evolution can claim comprehensiveness unless it answers that question.

The unique perfectness of cancer selection as the mechanism that made animal-manufacturing possible can be appreciated if we compare quality control in nature to the way it works in the mass-manufacture of the things humans produce.

I have observed at close hand the mass-production of products as varied as electronic control devices, pharmaceuticals, surgical sutures, pressure-sensitive tape, hospital bandages, telephone cable, copper tubing, brass fixtures, nails, chain link fences, aluminum foil, textiles, porcelain plumbing fixtures, lawn furniture, gasoline, soap, cosmetics, and probably a few other products I no longer recall. In every one of those facilities the people responsible for production installed and supervised procedures that controlled--regulated and verified--the quality of the end products.   The actual control methods implemented depended on the nature of the product manufactured and on the procedures employed in the process. Some useful generalizations can be made, however, about all of them and about all manufacturing operations.

In every case management of the factory established specifications of the products. These were in writing and always included *precise* descriptions of the following:

1. Raw materials.
2. Parts and sub-assemblies.
3. The finished product.

Other information, such as the machines to be used in manufacture and procedures to be followed by the workers might also be codified.   However, precise specifications of the *materials* and the *parts* that comprise the end product are essential to any effective quality control procedure, and, since by definition successful manufacturing is quality-intensive, they are essential to the operation of *any* manufacturing system.

In practice, raw material is routinely tested to ensure that it meets the specifications.   In a factory where pharmaceuticals were manufactured, quality control personnel routinely tested batches of newly arrived chemical raw materials in a gas spectrometer. This instrument, in which materials are heated until they reach a gaseous state, provided a precisely measured profile of the chemical makeup of the material placed in the instrument.   By comparing the profile of tested chemicals to a master-profile the quality control technicians could determine whether or not the incoming material met the established standards.

In addition to raw material verification, in any well-run plant quality control tests are made throughout the manufacturing process. And once all manufacturing is completed the finished goods are inspected to be sure that the end product meets specifications. In general, the complexity of the product determines the intensity of the quality control procedures. Products that demand more precision in manufacture require more stringent controls than those made less precisely. The manufacture of high performance aircraft demands more precision than the production of common nails. Quality control is more intensive and less tolerant of deviations from the standard in aircraft manufacturing plants than it is in nail factories.

Regardless of the relative complexity of the product or the procedures, however, all quality control systems share two general characteristics in common: they all depend on human *intelligence* and they all use, to some extent, *sampling*.

The need for human intelligence is obvious. Even if quality verification is automated, human beings establish the standards, program the computers and monitor the instruments.

Sampling is also essential, for two reasons.

First of all, many quality control procedures are destructive. Material tested in a spectrometer is evaporated and can't be recovered. Metal parts subjected to stress tests are destroyed in the process. Obviously, destructive testing can be performed only on samples.

The other reason for sampling is economic. Although weighing a loaf of bread doesn't destroy it, the people who run industrial bakeries might justifiably weigh only one loaf out of every thousand (or even fewer). Spot checking is commonsensical where manufacturing procedures are uniform and automated, and where experience shows that it is virtually as effective as examining every part and every finished product.

Now let's look at the problems faced by nature's manufacturers. If human intelligence and sampling are essential elements in all human-designed manufacturing systems how could nature have succeeded without them? We must respond fully to that question for nature most assuredly had no access to either. Yet without any intelligence and without using sampling techniques nature far outstripped man's ability to manufacture; millions of animal gene pools routinely produce, in great numbers and with mind-boggling efficiency, the most complex multi-part objects imaginable. Those brilliant results imply brilliant causes: powerful, evolutionarily effective, quality control mechanisms *must* have been at work. But what were they? How did they function and how did they manage to keep pace with the increasing demands for precision as evermore complex animals evolved?

Logic dictates that a plausible solution to the problem of nature's manufacturers must meet these criteria:

Because sampling was impossible, *every* part (cell) of *every* product (animal) had to be monitored to ensure compliance with the master specifications.

Operating without the intervention of intelligent beings, the control mechanisms had to be automatic.

The quality control system must *itself* have caused the gene pools to change over time *in ways that favored the system's evolution to its present level of efficacy.**

The only mechanism that meets those criteria is cancer selection. It satisfies all of them:

Oncogenes monitored the manufacture of every cell. They gave genetic material inside cells the same ability to communicate with the gene pool that only whole organisms possessed in non-animal lineages.

Initiation and execution was automatic. Cancer went into

---

*This statement is one of the most compelling arguments in the book.

action when the manufacturing process deviated from the specifications.

Every time cancer killed a developing animal it *irrevocably*
biased the evolution of that gene pool in favor of precise manufacture.

Because of cumulative effects of purgation by cancer selection,
all animal gene pools acquired the ability to mass-manufacture with
great efficiency. Oncogenes forced the selection of anti-oncogenes
and all anti-oncogenes enhanced precision in self-manufacture.

The readily observable ability of nature's dazzlingly efficient
manufacturers (all extant animal gene pools) to mass produce--
faultlessly--trillions of complex organisms informs us that only selection pressure from cancer could have done it. To consider any
other explanation as *possibly* valid is bad logic, and bad logic is bad
biology.

All human-directed manufacturing systems depend on feedback. It
is not enough for the quality control mechanisms in a factory to
detect serious and continuing problems on the factory floor. The
information must flow. Industrial engineers and others responsible
for the operation, who might work at some distance from the
problem, must learn of the difficulties in order to correct them.

I have already explained how natural feedback changed the
population of *Biston betularia* from grey to black moths. But that
type of feedback--the central mechanism of natural selection and of
neo-Darwinism--originated with whole organisms. Cancer selection
introduces the concept of feedback from the lowest level of biological complexity to the highest; from molecules to gene pools. My
theory says that production of each new somatic cell in a developing animal--the basic, repetitive step of organism-manufacture--was,
thanks to cancer, an evolutionarily significant event.

With every cell containing triggers capable of terminating the
process--killing the juvenile--whenever the DNA instructions were

not implemented with exactness, animal genes were forced to seize command--they had no choice--over the smallest detail of the animals' construction. Unlike plants and the other nonanimal lineages, in animal gene pools *every* act of mitosis in developing animals was a potential threat to the very survival of the genes.

Feedback in developing plants and plant-like cell colonies was much weaker. *Nothing* that happened inside a single cell could kill the organism. Individual somatic cells were never subject to the "be-precise-or-else" imperative established by cancer selection. Molecules inside cells could never communicate directly with their gene pools.* Without the cancer mechanism, only events that affected the *entire* organism could feedback to the gene pools. Without a flow of information from individual cells gene pools of nonanimal multicells could never produce truly complex organisms.

Cancer selection's power to create complexity lies in its simplicity and in its universality. Unlike natural selection, it rendered no judgments on specific physiomorphological propositions contained in the zygote's DNA. It simply laid down, and rigorously enforced, a fundamental rule of animal life: the detailed how-to-manufacture-an-organism instructions the zygote received from its parents were to be implemented in all cells with great precision.

Its simplicity, power and universality enabled cancer selection to function in all the lineages that descended from those Pre-Cambrian worms. It played the identical role--central and essential--in the evolution of animals as grossly different from one another (and as distantly related to each other) as the vertebrates, the insects and the cephalopods.

It is obvious that just as cancer selection mandated precise cell construction by ruthlessly eliminating imprecise replicators, an ana-

---

*It is difficult to imagine how events inside an individual somatic cell could kill an organism not capable of dying of cancer--for example, an entire tree. I suppose that at the earliest stages of development when the entire body consists of a small number of cells, a gross mutation in one cell could have fatal consequences. If the zygote had just divided and both of the two daughter cells were dysfunctioning as the result of mutation, that single event (the mutation) could be fatal to the very young organism. But once the number of cells increased beyond a modest number (12? 24? 640?) the tree would have the potential to survive even with some dysfunctioning cells.

logous process acted at levels above the cell. Misreplicated and malfunctioning organs made up of perfect cells would be of no use to any animal. We can be certain that many juveniles died of organ-failure and organ system-failure caused by incorrect replication. But without extreme process-to-program fidelity at the molecular and cellular levels no organs or organ-systems could have existed.

The alternative to complete acceptance of my explanation for the evolution of animal-manufacturing systems is to remain imprisoned within a grievously impaired theory, a blatant intellectual hypocrisy that takes for granted the magnificent results of precise self-manufacture--*natural* selection, after all, could act only on animals produced by faultlessly implemented genetic programs--but rejects the only plausible mechanistic explanation for it.

Because my theory, unlike neo-Darwinism, explains the evolution of those systems it can, with justification, be called the *first* comprehensive theory of evolution.*

---

*Despite the title of his 1988 book, A Theory of Evolution of Development, Wallace Arthur does not even attempt to do what I do in this chapter. He, like the other theorists he cites (including Waddington and Goldschmidt), failed to see process-to-program fidelity as both a major theoretical challenge and the key to understanding the evolution of development.

Seven

Why Did They Bother?

The chicken is the egg's way of ensuring the production of
another egg.

            Samuel Butler

I cut off the heads of the one that had seven, and after a few
days I saw in it a prodigy scarcely inferior to the fabulous
Hydre of Lernaea. It acquired seven new heads...But here
is something more than the legend dared to invent: the seven
heads that I cut off from this Hydre, after being fed, became
perfect animals...*

            Abraham Trembley

According to evolutionist O. Ledyard Stebbins, multicellular
organisms evolved from unicells at least 17 times. Only one of
those origins, however, was of the complex animals, and their ap-
pearance--which was sudden--has long puzzled evolutionists. This
is what Harvard biologist Ernst Mayr said about it:

> the marvelous radiation of the invertebrates was indeed a
> comparatively "sudden" event in the late pre-Cambrian between
> 700 and 800 million years ago.

---

*Under the classification method used in this book, hydra are nonanimals.

--and this was his explanation for the great event:

> Presumably, a whole series of factors contributed to this outburst: There may have been a change in the chemistry of the oceans, [genetic changes] may have become more frequent, and there may have been changes in the ecosystem...Perhaps we will never know.

I disagree. My theory asserts that the outburst, known as the Cambrian Explosion, was not caused by a grab bag of mysterious factors. I say it was caused *solely* by the introduction of cancer selection.

I can explain that event--the very origin of animals--by concentrating the illuminating power of my theory on those characteristics that distinguish animals from the other multicells. In doing that I accomplish what the old theory does not even attempt.

I had the first glimmer of the idea that led me to develop this theory in September 1977 while reading *The Selfish Gene* by Richard Dawkins of Oxford University. In his book, Dawkins uses, to good effect, inferences drawn from the observation that although all organisms die, the genetic material in all existing lineages has, so far, been immortal. He portrays the immortal genes as selfish "replicators" who use organisms as disposable survival machines or "vehicles." In all past generations the vehicles' function was to breed--so that the replicators could continue to survive inside subsequent generations of vehicles. In this analytical scheme, all organisms had the same simple job: to serve as temporary hosts of the genes in order to ensure the survival of those immortal replicators.

Dawkins uses this genes'-eye view of evolution to explain a paradox: why do organisms sometimes behave in ways that appear not to be in their own best interest? As an example, he describes certain mother-birds who risk their own lives when a predator, such as a fox, approaches the nest and their helpless young offspring. When she spies the fox the mother becomes an actress. She lowers one wing, as if it were broken, and wobbles away from the nest. Her hope is that the fox, seeing her convincing imitation of a wounded full-sized bird, will be tempted to attack her instead of her hatchlings. Dawkins argues--and I found him convincing--that the bird behaves in this near-suicidal way because she is manipulated by her genes. If the fox falls for the performance, he turns away from the vulnerable nestlings and attacks her. Then, at the last minute, the mother escapes the fox's snapping jaws by flying off on her "broken" wing.

Dawkins asks why doesn't the mother save her own neck by simply abandoning her offspring to the fox? His answer is that by putting herself in peril the mother acts in a way most beneficial to her genes, including those copies of her genes inside her offspring. The mother's behavior is caused by genes "for" her potentially suicidal behavior. Genes that cause her to behave that way were selected because they enhanced *their own* survival.

Dawkins' perception of the relationship of animals and genes, which at first seems both cynical and overly abstract, is a powerful explicator of certain animal behavior. I also found it useful in developing my own theory. It helped me to shed light on events in the distant past. I isolated characters that require explanation--which is not as simple as it might seem--and then asked the difficult but potentially enlightening question: How did those characters enhance the survival of primordial genes?

Dawkins' view* of the relationship of genes to organisms--*replicators using vehicles*--suggests another idea:

> *All* genes (replicators) in *all* lineages, animal or nonanimal, that now exist were *equally* successful. Every surviving lineage has produced, for hundreds of millions of years, an unbroken chain of sufficient vehicles.

It is inarguable: if the genes in an individual organism survived and moved into the next generation of offspring in numbers sufficient to ensure the genes' survival then their journey in that particular temporary host had been a *complete* success. That was it. As

---

*As Dawkins acknowledges, this idea was anticipated by the pioneering theorist, August Weismann, who, long before genes were discovered, developed the idea of "the continuity of the germ plasm."

far as the genes were concerned *nothing else about the organism mattered.* All genes that managed to create *satisfactory* survival vehicles for hundreds of millions of years were winners. Gene pool survival was the archetypical "pass-or-fail" test. No bonuses or extra credits were earned by lineages that created complex--*interesting*--organisms.Like most people, I find jellyfish boring. But genes have no obligation to impress us, they function under no imperative to enchant. They are neither artists nor entertainers. Genes exist solely to survive by replication, and in the great game of survival, jellyfish--simple and boring though they are--have been every bit as successful as human beings.

That brings us to the question that heads this chapter. If jellyfish, sponges and plants were sufficient, what on earth drove the genes of the earliest animals, *and only those genes,* to produce organisms that were strikingly *different* from the cell colonies?

I've already identified the most profound difference--the great complexity of the animals. Using the replicator-vehicle model, both nature--in the form of the far simpler plants and other nonanimals--and logic--under the law of parsimony--inform us that the genes ought to have followed the most economic strategy for survival. They ought to have chosen the simplest vehicles. Tissue-level bodies without vital organs worked for plant and jellyfish genes, but contrary to neo-Darwinian expectations, they were not good enough for animal genes.

The origin of complexity is related to another basic characteristic that distinguishes the animals from nonanimals: the uniformity of animal phenotypes within species. In contrast, nonanimals exhibit great phenotypic plasticity. No two maple trees, out of all the trillions that ever existed, had precisely the same arrangement of branches, leaves and roots--not to mention individual cells. The DNA in charge of maple tree construction does not mandate rigid pre-determination of individual bodies. This *laissez-faire* approach to development predominates throughout the plant kingdom and in sponges and cnidarians. Among hydra (cnidarians), individuals in the same species may even have different numbers of "arms." (They're actually simple tentacles made of transparent, jelly-like tissue.) Yet real animals, be they insects, mammals or octopuses, are all made with very little variation among conspecifics. In some species of nematodes even the cells are arrayed in hyperuniform fashion; identical numbers of cells (in *Caenorhabditis elegans* its exactly 959) are lined up in precisely identical arrangements. Biologists, who offer no explanation for the origin of that extraordinary uniformity, have at least given us a name for it: *eutely.*

Neo-Darwinism ignores organismic similarity. It emphasizes *differences* among individuals. According to that theory, natural selection "worked" on the variations; it retained those that gave the individual animal a reproductive advantage. Neo-Darwinists are right to treat differences as important; variations and their selection were essential to evolution. But that is no reason for theorists to close their eyes to the reality of dramatic uniformity of animal bodies. Scientific theories, after all, have only one purpose: to explain natural phenomena. In developing a theory of origins the more pronounced phenomenon--intraspecific uniformity of animals' bodies--ought to attract more attention than the minor variations.*

Another characteristic common to all animals and therefore of very early origin is the lateral symmetry of their bodies. Although some nonanimals might achieve rough radial symmetry in adult form, among *all* animals strict bilateral symmetry is present at the early larval stages of body construction, even in species where adults are not symmetrical. (Bivalve mollusks--oysters, clams and the like--and echinoderms, starfish, tend not to be laterally symmetrical in the adult form, but they display it as embryos.) Why did only** the earliest animal genes select that unique body plan?

Of course, the bilateral form enabled *later* animals to move with skill, speed and grace. But the symmetry came first, long before there were any swimmers, runners or flyers. Nature could never anticipate future needs or conveniences; no serious old-theory

---

*We humans would probably be more aware of our similarities if we had transparent skin. Most of our individual dissimilarities are located on our highly visible exteriors (skin, hair, facial features, and so forth) but our most complex parts, the internal organs, vary little among individuals, even those who are distantly related. African bushmen, American Indians and Australian aborigines probably have not had a common ancestor for hundreds of thousands of years, but a team of European and Asian surgeons could perform complicated procedures on all of them; there would be no noticeable differences in any organ.

**I comment on the--quite different--symmetry found in some plants and other nonanimals in Chapter Nine.

proponent believes that symmetry was selected in the Pre-Cambrian for the benefit of modern animals. So the question remains: If nonsymmetry or simple radial symmetry was good enough for the nonanimals, why did the earliest animal genes *bother* with the more precise--and therefore more demanding of exact, genetically-controlled development--laterally symmetrical arrangement?

Other fundamental differences between animals and nonanimals, all of them concerning the relationship of morphology to reproduction and regeneration, cry out for explanation. Consider the following comparisons of characteristics of animals and nonanimals:

	Animals	**Nonanimals**
Multiple sex organs?	*Rare*	*Prevalent*
Slowing of somatic cell production after reproducing (aging)?	*Yes*	*No*
Regeneration of the body from small pieces?	*Rare*	*Prevalent*

All the features in the nonanimal column had obvious selection benefits in terms of the old theory. They are fully explained by natural selection. *But the old theory explains nothing in the animal column.*

Multiple sex organs, found in virtually all plants, cnidarians and sponges, permit organisms to reproduce potentially greater numbers of offspring. Although some animals (queen termites, for example) have multiple sex organs, most get by with a single organ. But if multiple sex organs offer many obvious advantages to the genes, why weren't they selected early in animal evolution?

And why didn't the animal genes, right from the outset, create organisms that lived--and reproduced--much longer? A coral reef, which originates as a small multicell that grows vegetatively to an enormous size, can be considered a single organism. The reefs' normal mode of death is similar to that of trees; they collapse from excess mass. But before that happens the organism lives a long life--as long as 800 years. Other cnidarians live for extraordinarily long periods. In Victorian England, some families kept sea anemones alive in their parlors for *ninety years*! Unlike goldfish and other modern home aquarium favorites, these creatures, which were kept in large jars, died not from old age, but only if they were neglected or if someone accidentally knocked over the glass vessels. Modern specialists have determined that the anemones and other long-lived cnidarians never display signs of aging. According to Bernard Strehler:

> in the opinion of most recent investigators, certain Hydrazoa, such as hydras and probably many species of Anthozoa [coral] do not undergo individual aging...

The sponge, another simple multicell that has undergone little noticeable evolutionary change, can live for fifty years.

World champions of longevity are found among the plants. Redwoods, giant sequoias, white oaks and other trees live for centuries. But even they are surpassed by the most spectacularly long-lived of all organisms, the huckleberry. Although they appear as individual bushes on the surface, huckleberry plants are actually enormous single organisms connected by underground root systems. Princeton Biologist John Tyler Bonner describes one specimen that is 2000 meters in diameter. Its estimated age is 13,000 years!

Spectacular longevity, which makes a great deal of sense in neo-Darwinian terms, occurs in organisms that lack vital organs and in lineages that have undergone relatively little evolutionary change for hundreds of millions of years. In animals, however, the patterns are dramatically different. They all *age*. And the evidence says animal senescence is controlled by genes.

O.M. Pereira-Smith and J.R. Smith have determined that "limited proliferation [of somatic cells] is a result of a rigorously programmed series of events."

Thomas E. Johnson, a geneticist at the University of Colorado, recently identified a gene in nematodes (*Caenorhabditis elegans*) that actually shortens life. He demonstrated its function by mutating that gene with the result that the maximum life of these tiny worms increased from 23 days to 37 days. The obvious conclusion: in its normal (unmutated) state the gene decreases life span by 38%.

Nematodes are perhaps the most primitive of all animals. This suggests that their gene "for aging" was selected very soon after, or even concurrent with, the origin of animals. But why? Genes that deliberately shorten organisms' lives defy neo-Darwinian explanation.

Then there is the baffling problem of regeneration. Why didn't all the animal gene pools select and retain genes for the tactic, used by plants and many cnidarians, of regenerating entire organisms from small bits? The plants' ability to regenerate is known to all. Other nonanimal multicells also regenerate prodigiously. When experimenters cut an individual hydra into 200 small pieces each piece grows into an entirely new organism, a degree of regeneration with obvious survival advantages. If a predator left a few pieces of a single hydra floating in the sea, regeneration would actually *increase* the population of gene-bearing organisms!

If ever there existed a physiological device that is most satisfyingly explained by the old theory of evolution it is that ability to regenerate entire organisms from small pieces. But despite its obvious advantages, only a few animals come close to the nonanimals in regeneration ability. Two species of annelids, polychaetes *Ctenodrilus* and *Dodecaceria*, can regenerate entire worm-bodies from a single segment. Other worms can regenerate their anterior ends and some can regenerate their tails. Salamanders can replace tails. Echinoderms like starfish can regenerate lost arms, and entire bodies from one arm. But the more organized animals cannot regenerate nearly as well. None of the insects or other adult arthropods and none of the vertebrates regenerate flamboyantly. They added many survival-enhancing characters during evolution yet declined to select or managed to discard (if, as seems more likely, their ancestors possessed it) that simple, obviously beneficial, survival technique.

Perhaps Andrew Trembley's amazement (which I quote at the beginning of this chapter) when he discovered the extent of hydras' regeneration capability is understandable. But all animals have the same "ace in the hole" that hydra possess. They too have a full complement of the DNA needed to clone an entire organism in the nucleus of every cell. What should astonish us is not the hydra's regeneration capability--it makes neo-Darwinian sense--but that most animals *lack* something natural selection ought to have selected and preserved.

Unless, of course, the old theory of evolution is wrong.

The old theorists' explanation for these puzzling differences between animals and nonanimals? They *ignore* them! They want it both ways. Faced with a senseless characteristic--genetically-controlled senescence is the best example--in one group and its sensible absence in another--no aging in nonanimals--they ignore the fatal consequences to their theory of the conflicting evidence. M.R. Rose, for example, in a comprehensive 1985 survey of theoretical work on senescence, mentions several scientists, including the late Nobel laureate Peter Medawar, who have confected natural selection "explanations" for its origin in animals. But he fails to identify anyone who also explains why plants do *not* age. Such pseudo-explanations are unworthy of serious attention.

Karl R. Popper, the philosopher of science, decided that the theory of evolution by natural selection is scientific even though evolution itself could not be reproduced in a laboratory.

The theory of [evolution by] natural selection is a historical one: it constructs a situation and shows that, given that situation, these things whose existence we wish to explain are indeed likely to happen.

Although Popper is correct to conclude that a theory of evolution can be scientific, the old theory fails to meet his own criterion. The situation it constructs--which emphatically excludes cancer selection in animals--does *not* explain "things we wish to explain." Without cancer selection, the genes in those primordial lineages simply would *not* have produced animals. There was no imaginable reason for them to have *bothered.* But those theoretically preposterous creatures exist and the message they deliver is emphatic: the theory that says they ought not to exist is wrong.

Because I establish an entirely *different* situation, or set of postulates, my theory explains what the old theory takes for granted: all the fundamental features of animals. It says that once cellular death-triggers (functioning oncogenes) were embedded in the animals' genetic programs, developing animals that did not avoid cancer were killed; their genes were extinguished. Cancer selection worked *against* creation of *excessive* somatic cells and *for* a consortium of mechanisms that minimized the potentially lethal somatic cells in prereproductive animals. What follows are my theory's explanations for the origin of characteristics that distinguish animals from other multicells.

Uniformity of morphology. We know from human experience that repetition is an efficient learning mechanism. That explains phenotypic uniformity. The construction of the same body model over and over again for countless generations reduced the possibility of error in the production of cells. Phenotypic uniformity was adopted as a cancer avoidance mechanism.

The importance of strict morphological uniformity in the origin of biological complexity cannot be exaggerated. As a few moments of thought will make clear, in a system operating without the intervention of an intelligent being complexity could not have evolved unless the system first learned to produce objects that were very much *like* one another. Analogously, human-engineered manufacturing systems that have been automated the most (those that use machines operating with minimal direct human supervision) are those that produce identical products in large volumes. Ball point pens, bottled soft drinks, and Band-Aids are a good examples. At the other extreme, man-made products produced with great variation from one another--custom-built furniture, architect-designed houses and haute couture garments--require intensive human involvement in every step of manufacture.

Animals are far more complex than *any* objects crafted by human intelligence. The system that created them, characterized by automatic feedback and self-correcting gene pools and the complete absence of supervising intelligence, could not possibly manufacture end-products that varied much from one another. The phenotypic uniformity mandated by cancer selection was essential to the evolution of complexity.

Lateral symmetry. The error-reduction benefits of repetition also explain the selection of genes for bilateral symmetry. Each side of a bilateral animal is a mirror image of the other. In development, repetition of the same plan on both sides of the longitudinal axis simplified construction and reduced cancer risk.

Moreover, bilaterally symmetrical body plans permitted the structure of very *small* animals. In some modern animals, such as mites and nematodes, specimens are actually smaller than many single-celled creatures. Smaller animals require fewer cells than larger animals and are thus less likely to incur lethal cancer.

Single sex organs. Although multiple sex organs have the
potential to create more zygotes, their existence depends on greater
numbers of somatic cells. A three-hundred-year old white oak will
produce many thousands of acorns every year, but in order to make
that many acorns the tree must produce trillions of new somatic
cells in branches, leaves and flowers. In animal lineages, because
the genes operated under the oncogene-in-every-cell imperative, the
creation of each cell was a potentially lethal undertaking. Cancer
selection mandated minimal numbers of somatic cells. That meant
single, not multiple, sex organs.

**Reduced somatic cell production once sexual maturity is
reached.** Without oncogenes in each cell the plant and other nonanimal genes built simple organisms, which, if permitted by environmental conditions, lived for long periods, vigorously producing
gametes (with those multiple sex organs) to the very end of their
lives.

The animal genes could not adopt that "open" approach to development; animals that grew without the strict discipline over cell
production incurred cancer. They and their genes perished. That
led cancer selection to impose what I call the "go and stop" strategy. Once the animal was complete and ready to mate its genes
could *go* about the task of constructing brand new bodies (the next
generation) using a tested development program-process that was
free of lethal cancer. The parent's somatic cell production machin-
ery--at worst a potential threat to the genes, at best no longer
needed--would *stop*.

Although we cannot return to the Pre-Cambrian era and in-
spect the earliest animals to verify this idea, we can consider a
present-day animal that seems to follow a clear "go and stop"
strategy.

The nematode *Caenorhabditis briggae* is very small: it consists
of less than 1000 cells. It matures in five days, stops reproducing
in 12 to 14 days, and dies in 25 to 28 days. It's genes, in other
words, create the smallest possible survival vehicle for themselves,
and, having (likely) accomplished the genes' objective of creating
offspring, promptly kill it, probably by activation of that gene
discovered by Johnson. Severe selection pressure for features that

minimized the likelihood of prereproductive cancer explain these mechanisms.

My theory thus explains the origin and function of animal *aging*. The programmed shutdown of the cell-renewal process was one of several mechanisms selected to avoid cancer in the earliest animals. It also explains the absence of senescence in plants: they don't get cancer. (Later, in some animal lineages, including our own, the onset of aging was delayed. I explain how that happened in Chapter Nine.)

Avoidance of flamboyant regeneration. Aggressive regeneration obviously requires increased somatic cell production and heightened cancer risk. Selection purged most lineages that attempted it.

Significantly, modern research findings, which I review in Chapter Eleven, show that excessive regeneration can lead to cancer in humans.

There is another powerful reason to conclude that these fundamental properties of animals came to exist as a result of cancer selection.

Biologists have long known that not long after their origin the bilaterally symmetrical animals divided into two major groups. One group is the *Platyhelminthomorpha*, animals with only one opening to the gut. All the other animals have two openings, a mouth and an anus, a group named (by biologist Michael T. Ghiselin) the *Proctozoa*. Among *Proctozoa* are animals as different from each other as insects and vertebrates, animals that have not had a common ancestor for more than 600 million years.

That history is crucial to evaluating the two competing ideas--evolution by natural selection and evolution by cancer selection--because modern representatives of those ancient lineages, the platyhelminths (flat worms) on the one hand, and the *Proctozoa* on the other all share those fundamental characteristics I focus on in this chapter:

Their DNA rigidly controls development by producing phenotypes with extreme conspecific uniformity, precisely arranged cells and somatic complexity; all have vital organs. Their bodies are laterally symmetrical, they age under the control of a genetically-determined program and most do not regenerate severely damaged tissue.

The theoretical choices that might explain those features are as follows:

Natural selection. Each property came to exist *and was retained in each lineage* in response to numerous--as yet *unidentified!*--selection agents.

Cancer selection. Each property came to exist *and was retained in each lineage* in response to an identical threat to all the gene pools: lethal cancer in juveniles.

As I've already made clear, the idea that natural selection could account for both the presence of those properties in animals and their absence in nonanimals is simply not credible. But the early divergence of the three animal groups (platyhelminths, arthropods and vertebrates) places an *additional* burden--an utterly unbearable one--on the narrow shoulders of natural selection. It is several orders of magnitude more unlikely--so unlikely as to be absurd--that natural selection, caused *multiple* origins and *multiple* retentions of those characters in response to... Well, in response to whatever melange of scenarios neo-Darwinists might possibly concoct for *each* separate line of descent.

My theory eliminates the absurdity of multiple causes for origins and retentions of identical features stemming from unidentified and illogical (because the characters are absent in nonanimals) selection agents. Although I think it probable that those properties all originated prior to the great ancient divergences, their retention in each of the lineages came afterwards. My theory, because it postulates a *common* cause for all of them, is not discomforted by that fact.

Those who remain faithful to the old theory will not appreciate being told that they embrace logical absurdity. They should not, however, think me rude for doing so, for I am merely pointing out the consequences of their own intellectual decisions.

I could have solved the mystery of the origin of those fundamental animal features by using the "cause of death" rule developed in Chapter Four: "What caused the death of young animals that lacked these fundamental characters?" As with the "Why did they bother?" approach, this line of reasoning leads inexorably to the conclusion that only the ruthless elimination by cancer selection of animals that did *not* possess them can explain those properties' origin in (and retention by) all animal lineages.

It is well that the old theory's adherents are, for the most part, silent on the origin of these features for they cannot be explained by natural selection. And failure to explain them is failure to explain the origin of animals. My theory seems to accomplish that task.

Eight

Killer Sunlight: The Sea Animals

> The causal relationship of historical phenomena ordinarily
> must rely on inferences from observations.
>
> Ernst Mayr

We can penetrate deep time and uncover the past's most important secrets by carefully analyzing characteristics of modern and fossilized animals. Properly interpreted facts tell us how great numbers of nonbreeders died. Using the same logic, I have concluded that *sunlight-induced* cancer was a major killer of early animals.

Before explaining how I arrived at that conclusion let us first consider the fossils of early plants and other nonanimals. Those organisms not only did not avoid sunlight many of them were attracted to it. Green plants could not survive without it. Many cnidarians floated on the oceans' surface without protection. Other cnidarians, and the sponges, lived at relatively shallow depths, their body cells unshielded from sunlight, which can penetrate water to 150 meters. Some coral lived, as now, inside exoskeletons that provided some diminution of sunlight exposure, but even the stony corals were shallow-water creatures and grew heliotropically, toward the sun. In short, the fossil evidence shows that most nonanimals always acted as if sunlight could never harm them.

But the animal fossils tell an entirely different story. They reveal that for the first 300 to 400 million years, when all animals

But the animal fossils tell an entirely different story. They reveal that for the first 300 to 400 million years, when all animals lived in the sea, *all somatic cells were sheltered from exposure to sunlight, even from sunlight filtered through sea water.* Moreover, the further back we go in time, the more primitive and cumbersome were the animals' sunlight-avoidance mechanisms. According to James W. Valentine, Professor of Geologic Sciences at the University of California:

> The earliest animal fossils are burrows that begin to appear in rocks younger than 700 million years, late in the Precambrian era. Both long horizontal burrows and short vertical ones are found, comparable in size to the burrows of many modern marine organisms. The ability to burrow implies that the animals had evolved hydrostatic skeletons, that is, fluid-filled body spaces that work against muscles, so that the animal could dig in the sea bed. Although...simple... sea anemones manage...to burrow weakly, long horizontal burrows suggest a[n]...active animal...*This is quite an advanced grade of organization to find near the base of the fossil record*...[Emphasis added.]

Those earliest animal fossils strongly support the idea that many juvenile animals died of sunlight-induced cancer. Few habitats offer more shelter from sunlight than burrows in the sea bottom. Applying the "cause of death" rule, we can infer that animals that did not live in those burrows died *for that reason.* But what killed them?

We can dismiss the possibility of predators. Animals that devour other animals tend to be large and complex. Big predators have left a lot of fossils, but not in the Pre-Cambrian. The hiders were there long before the seekers.

Since, as Valentine points out, burrowing demanded complexity we can ask another question: Why did the worms' genes bother to produce animals complex enough to bury themselves in the sea bottom? Because I have already argued that high-level complexity emerged only in organisms that were subjected to intense cancer selection, the most plausible explanation for burrowing worms is that their gene pools were purged by *sunlight-initiated* cancer. Cancer selection explains both complexity and burrowing. Occam's Razor--the rule of parsimony--tells us that solutions that solve more than one problem are logically superior to any hodgepodge of unconnected proposals.

Alternative explanations for the complete sheltering of somatic cells and the emergence of complex organisms--phenomena that did not occur, remember, in other multicells--involve mysterious predators and the miraculous appearance (that's the only thing to call it) of locomotion systems, circulatory systems, central nervous systems and complex, light-sensing organs. All those complexities would have been needed in an animal that burrowed.

According to Valentine, those burrowing worms lived in the mud and sand at the bottom of the sea an astonishingly long time--120 million years! And when some of their descendants emerged from living *in* the sea floor, they did so in a way that strongly suggests sunlight continued to kill. The creatures simply moved upward to the sea floor itself where the radiation-shielding sea water remained between them and the killing rays. And all those bottom-crawling, complex animals (all of them more complex than the worms) had thick external hard parts--cumbersome, heavy, noncellular outer coverings.

Why did the genes bother with such elaborate devices? Heavy armor puts an enormous burden on the resources of the animals. They had to devote developmental effort to building it and then they had lug it around for the rest of their lives. My answer, of course, is that those coverings protected the animals from the same threat that drove their ancestors into the mud. Because ultra-violet radiation could kill an animal if it struck just one cell, the early animals' gene pools adopted a cautious low-risk strategy: zero cells were exposed. All those descendants of the burrowing worms, and all their mitotic cells, lived inside shells or behind chitinous scales and exoskeletons.

When the earliest vertebrates appeared they, too, arrived wearing suits of armor. The earliest fish (*ostroderms* and *placoderms*) were covered with thick plates, precursors of the scales of modern fish.

Significantly, the armor plates that encased the early fish and the other armored animals were heaviest at the top, an arrangement still evident in existing armored animals. My theory explains this positioning: the thicker armor faces the sun.

(An alternative idea, that external coverings emerged as protection against predators, appears in more than one biology textbook. But that proposal doesn't address the problem of heavy protection on top and lighter protection on the bottom. And it is a more serious problem than might appear at first glance, for the selected arrangement actually weakened protection against predators. A predator could crush the thin armor against the heavy armor, which would act as an anvil. Common sense says that if external hard parts served primarily as predation-protection devices they would have been vertically symmetrical, as thick on the lower part of the body as on the upper. Then they would have been as difficult to crack as coconuts. Alas for natural selectionists, the armor is not equally thick all-around.)

The fossils show that the vertebrate fish eventually adopted much lighter coverings. In many of *their* descendants, including many terrestrial vertebrates, armor was dropped completely. But any newly exposed flesh continued to be protected against ultraviolet radiation. Virtually all exposed skin was heavily pigmented, and pigmentation is a known cancer-protective.

This pattern of gradual reduction of external outer coverings occurred in many lineages. Sharks, which may have some as yet undiscovered protection against cancer, have eliminated external hard parts. But they have retained pigmentation, and, as is the case with all pigmented animals both in the sea and on land, the heaviest pigmentation is on that part of the body most exposed to sunlight, further pointing to sunlight-caused cancer as the selection agent. (I comment further on the relationship of cancer to shark evolution in Chapter Nine.)

Like the shark, the octopus and the squid dispensed with exter-

nal hard parts. These mollusks, called *cephalopods*, descended
from mollusks that were heavily armored--and much simpler. Octopus and squid are probably the most transformed, and perhaps
the most complex of all the animals that never left the sea.* They
fit the pattern that my theory establishes. As the animals became
more complex they dropped the "blunt instrument" approach to cancer defense. Like many vertebrates they lost their external hard
parts. Logic tells us they could not have discarded the hard outer
shells unless the effectiveness of *inner* defenses against cancer were
strengthened. Modern research supports the logic. Bacq and Alexander, in *Fundamentals of Radiobiology,* report that an octopus
gland produces hydroxytryptamine, the "most effective radiation
protective agent known to exist."

Because they disposed of hard outer coverings, octopus and
squid benefited from more revolutions of the engine of transformation. Increased cancer selection pressure caused the lineages
to retain new anti-cancer genes. These anti-oncogenes enabled
them to build more complex bodies, which, in turn, increased
cancer rates. Cancer selection, the accelerant of transformation,
drove them to levels of complexity much higher than those of mollusks that remained inside protective shells--clams, oysters and the
like.

Other modern sea animals have even discarded pigmentation
and are nearly as transparent as jellyfish. A few fish species and
virtually all arrowworms (*Chaetognatha,* small, free-swimming animals found mainly in the Pacific) are transparent. Although trans-
parency and complexity in the same animal would seem to contradict my theory, these creatures rigorously support it.

All transparent sea animals avoid sunlight. Most are daily
vertical migrators. They swim downward--away from the sun's rays-
-in the daytime, and up toward the surface at night. Transparent
pearlfish spend their days inside the respiratory systems of sea
cucumbers and leave that protection to forage only at night. Some

---

* The marine mammals (such as whales, seals and sea lions) all descended from
animals that spent millions of years on land.

fresh water fish have become troglobitic; they inhabit rivers and pools inside caves. These fish have become transparent (and blind) since moving to the caves. Because the caves provide complete protection they do not need external morphological cancer defenses like pigmentation.

Interestingly, the arrowworms offer an additional confirmation of my theory. Although most arrowworms live in the deep ocean one species lives near the shore and cannot avoid exposure. Since they cannot possibly migrate vertically, members of this arrowworm species would seem to refute the theory. Refutation is emphatically denied, however, for this species of arrowworms, unlike all the species that live in deeper waters, is *not* transparent. It is pigmented.

By applying the "cause of death" rule to the mechanisms for avoiding exposure of unpigmented cells to sunlight used by marine animals we are led, once again, to the conclusion that cancer was the cause of an astronomical number of deaths of juveniles.

Common sense tells us that the earliest marine animals could not avoid light unless they could *detect* it. Cancer selection thus explains the origin and function of the first primitive eyes.

According to the Ernst Mayr photoreceptors or eyes evolved independently at least 40 times in the animal kingdom. That tells us there was intense selection pressure on all early animal lineages: the ability to detect light *must* have been a matter of life or death.

Some evolutionists, incautiously (and, I'm afraid, typically) carried away by what they can observe, are convinced that the original function of eyes was to see *things*, to avoid predators or to locate food. Those assertions ignore the nonanimal lineages including creatures like Portuguese-Men-Of-War and other jellyfish that are both predators and prey (sea turtles eat jellyfish) and whose lineages have survived very nicely for many hundreds of millions of years without keen vision and without a brain capable of processing the kind of precise information that a highly developed eye would capture.

Unlike the predator-prey hypothesis, my explanation for the

origin of eyes--that animals had to detect light in order to *avoid* it--explains the prevalence of acutely sensitive and complex light-detecting organs in the animals, and their absence in nonanimals.* The rule of parsimony tells us that the concurrence in the same early animals of sophisticated light detectors, extreme protection of somatic cells from sunlight and emergent somatic complexity all point to cancer selection as the common cause.

Despite that long history of sunlight avoidance by marine animals it is obvious that some of their descendants not only succeeded in the sun-drenched terrestrial habitats but became the most spectacularly complex of all animals. I explain how that happened in the next chapter.

---

*This idea also explains another long-standing problem with the old theory. Critics of natural selection, starting with H. St. George Mivart, Darwin's contemporary, have asked of what possible use would a primitive eye have been? They argued that (as Darwin himself had conceded) the ability to actually see objects clearly would have evolved slowly, a crude device would have been utterly useless. This is valid criticism and my theory eliminates it: if light itself could kill, even a crude detector would enhance survival.

Nine

Killer Sunlight: The Land Animals

I hold it to be true that pure thought can grasp reality, as the
ancients dreamed.

Albert Einstein

What can be explained by the assumption of fewer things is
vainly explained by the assumption of more things.

William of Occam

The theory says that animals that avoided sunlight the most
experienced less cancer and evolved the least. Earthworms, the
bivalve mollusks and other light-avoiders are of lineages that did
not undergo much transformation; all are relatively simple.

The logic, symmetry and power of the theory also tells us to
look for the *most* transformed creatures in lineages with the longest
histories of exposure to solar radiation. Ultra-violet light would
have intensified cancer selection and caused more revolutions of the
engine of transformation. Reality concurs with logic. The most
transformed of all animals are those whose ancestors left the more
sheltered marine environment to live on the sun-filled land surfaces
of our planet. Because greater transformation usually led to increased complexity, the most complex of all living creatures are
the surface-dwelling terrestrial animals and their marine descendants.

When judged by the number and variety of species the two
most successful invaders of land were the insects (and other arthropods such as millipedes and spiders) and the vertebrates. But

before explaining cancer selection's role in the evolution of those two enormous groups of strikingly different animals, I will comment briefly on two smaller groups of land animals, the mollusks and the annelids.

The common garden slug typifies land-based mollusks. That creature, a pest to gardeners, exhibits characteristics consistent with the heavy evolutionary influence of cancer selection. Although slugs crawl around without any shells or other obvious passive radiation protection devices, they crawl only at night. During daylight hours they hide beneath logs and other objects. That pattern of nocturnal exposure and daytime shelter is consistent with sunlight-induced cancer selection in the lineage.

Now let's consider *Oligochaeta,* the common earthworm. Like the transparent marine animals described in the last chapter, many earthworms are daily vertical migrators. They remain in their burrows during the day and crawl out only at night.

Oddly, the bodies of those night crawlers and other burrowing worms are *covered* with microscopically small eyes.

Why so many eyes? Vertebrates and insects, most of whom depend on acute vision for their very survival, manage quite well with just one pair of eyes. And among vertebrates that live underground, some, such as cave-dwelling fish and moles, have lost their vision; their eyes have atrophied. The disappearance of those fish and mole eyes is a stunning demonstration of how natural selection ought to work: change the environment so that the organ is no longer useful and the organ disappears.

So what about those earthworm eyes? No predators lurk underground. Visibility in burrows is near zero. If the eyes were there to detect any nocturnal surface predators the night-crawling worms might encounter, what could a slow-moving worm do to save itself if it saw a threatening animal? There are few phlegmatic predators. Predator-detection seems the unlikeliest explanation.

It seems, in fact, that if natural selection is our only analytical tool we should conclude that earthworms ought not to have *any* eyes. But their bodies are *covered* with them. What in the name of Darwin is going on?

The eyes make sense, of course, if they were created in response to cancer selection. If the function of worms' eyes is to detect, not predators, but *light,* having them all over the body makes a great deal of sense. If compound eyes convert light to pain they would encourage the animals to avoid it. That, of course, is exactly what those relatively simple, small-brained creatures do.

(As I explain below, although they too can get cancer, fish that live in caves and mammals that live underground fear the sun less than annelids. The disappearance of their eyes is also consistent with my theory.)

Previous explanations for their subterranean habitat selection and nocturnal activity (such as predator avoidance) fail to account for the annelids' bizarre compound eyes. They also fail to note that an entirely different group of terrestrial invertebrates, the arthropods, contain many species that spend their adult lives exposed to sunlight.

When they emerged from the radiation-shielding protection of the sea the founding lineages of insects (and other terrestrial arthropods such as spiders, millipedes and land crabs) had to strengthen their bodies' cancer defenses. They had no choice. Water blocks radiation. The land surface is far more carcinogenic.

Insect gene pools collected an arsenal of weapons. The two scientists who discovered insect cancer, in larval *Drosophila*, Elisabeth Gateff and Howard A. Schneiderman, identified several characteristics that they thought might explain the relative rareness of cancer in insects:*

(1). Unlike humans and other vertebrates, adult insects have few cells that divide. Cells that do not divide cannot possibly have cancerous offspring.

---

*Gateff and Schneiderman did not, however, suggest that these characteristics came into existence as a result of cancer selection.

(2). Most insects undergo metamorphosis during development and during metamorphosis the adult animal is created from "imaginal disc" cells, not from the division of somatic cells that comprise larval tissue. Any tumors that might exist in the larval tissue could be discarded.

(3). Much insect growth occurs without cell division; the cells simply get bigger and, in some cases, DNA replicates inside the cell. Again, unless they divide, no cells, not even those in which a major mutational event has taken place, will have cancerous offspring.

(4). Insect DNA is spectacularly good at replication. When normal (noncancerous) vertebrate cells are kept alive in laboratory vessels, abnormalities appear in DNA after a few cell divisions. Then all the cells die as the result (it is safe to presume this) of the genetically-controlled aging process. Some *in vitro* human cells, for example, cease to divide after 50 generations.* In contrast, observed insect cells divide more than *1,500* times without abnormalities. If DNA inside insect cells is more efficient at mitosis than vertebrate DNA it will experience less cancer. (This leads to the question of *why* vertebrate DNA is not as good as the insects in cell division. I explain that later in the chapter.)

Gateff and Schneiderman do not mention other insect characteristics that I think were selected as cancer defenders:

Insects shield their larvae from sunlight. The larval stage is the riskiest because cell division is at its peak. They place larvae in locations that afford heavy protection from sunlight exposure: underground, in mud nests, under the bark of trees, beneath rocks and inside hives.

---

*Human cancer cells can seemingly live forever in vitro. This shouldn't surprise us since it is their invincible capacity to divide that kills.

Insects are short-lived. *Drosophila*, the geneticist's favorite experimental animal, goes from egg to egg in ten days. They do so because the insect genes adopted the "go and stop" defense.

Another sign of "go and stop" cancer defense is the insects' small size. Because cancer starts in a single cell animals made up of few cells are less likely to experience a catastrophic cancer-causing event than those with more cells.

Notable exceptions to the small sizes of insects are the queens of the colonial insects--bees, termites and ants. However, those relatively large (and long-lived--some queen termites live for twenty years) insects spend most of their lives in heavily sheltered habitats; termite queens and ant queens live underground, queen bees in hives.

It is worth noting how one old theorist "explains" the insects' small size. Stephen Jay Gould mentions (1977) not one, but two "reasons." According to that Harvard paleontologist, the insects' breathing apparatus is appropriate for small animals, but not big ones, therefore insects are small. His other "explanation" is the insects' need to molt. When they shed their exoskeletons, the insects' bodies are soft. Large soft bodies would collapse, therefore the insects have small bodies.

Gould is, of course, using circular reasoning: Why is the breathing mechanism suitable for small animals? Because the animals are small. But why are the animals small? Because of the breathing apparatus!

I don't know why Gould stopped with just two. He could have listed lots of other "reasons" for the insects' small size: their tiny legs, diminutive wings, lilliputian digestive systems, wee brains, etc.

Gould's nonexplanations ignore the cancer-defensive features mentioned by Gateff and Schneiderman. He also doesn't explain the insects' short lives. He fails to recognize that insects have a panoply of mechanisms that reduce the possibility of cancer death through the simple, but highly effective means of sharply curtailing somatic cell production.

But if insects are tiny because they can get cancer why are humans and other vertebrates, all of which experience more cancer than insects, so much larger?

That's a good question. It gets a good answer.

The first terrestrial vertebrates, the ancestors of the largest and most complex land animals, left the sea armed with a unique anti-cancer weapon. This was something substantively different from the devices used by insects, or annelids, or mollusks. What those first terrestrial vertebrates had was a *second* line of defense against cancer, an array of mechanisms that (unlike cell-curtailment or sunlight-avoidance, which work by preventing the *initiation* of cancer) could kill cells *after* they became malignant. Those animals had a powerful *immunological system*.

The idea that a major function of the immune system, perhaps *the* major function, is to protect against cancer is not new. Robert A. Good and Joanne Finstad, for example, suggested in 1968 that "A primary *raison d'être* for the lymphoid system and certain immunities is surveillance against [cancer]." More recently, the catastrophe of AIDS, a virus-caused disorder that impairs the immune system, has provided strong empirical evidence of the system's vital role in fighting cancer. Many AIDS victims die of it. Other compelling evidence is the effort now underway in many major cancer centers, including the National Cancer Institute, to artificially enhance the immune system's ability to kill cancer cells. Researchers wouldn't undertake those programs unless they were convinced it already kills cancer cells.

All those facts support two related ideas of great importance to understanding vertebrate evolution:

Cancer *initiation* is a routine occurrence in vertebrates, the only animals known to have cancer-specific immune systems.

Surveillance and elimination of routine malignancies is a major function of the immune system.

Vertebrates had an *active* secondary defense against cancer. The evolutionary significance of the fact that it could only act against cells that were *already* cancerous is enormous. It cannot be exaggerated. The invertebrate lineages, which depended on single-phase passive defenses, did produce complex animals living in sunny habitats but the animals are either small and short-lived like insects or they are encumbered with heavy armor plate like land crabs. Other invertebrate terrestrial lineages produced simple animals like earthworms and garden slugs that hide from sunlight. But only the vertebrate lineages produced complex *and* large terrestrials, and, contrary to conventional interpretations, their back bones were not responsible. Immune systems and cancer did it.

Lots of cancer. The immunological system could not have emerged *unless* primary cancer defenses failed--repeatedly--to protect the animals. Evolutionary logic informs us that intense cancer selection was needed for complex cancer-specific immune systems to originate and to evolve to their present level of efficiency and complexity.

To get some sense of that complexity, consider the thymus. That gland, which is found in all vertebrates, attains its greatest size relative to the human body during the prenatal and neonatal periods. During that time it produces substances that activate certain genes inside lymphocytes, the killer cells that hunt down malignant cells and other antigens during the lifetime of the organism. After indoctrinating the body's T-cells--a process completed, in humans, about six months after birth--the thymus is turned off. It atrophies and eventually disappears; removal of the gland in adults does no harm. But without that good-for-a-lifetime processing of T-cells during infancy those cells could not function as cancer killers.

This is mind-boggling complexity. It's impossible to even imagine a man-made analog. The best I can do is a computer-driven police academy that not only teaches each cop how to identify and eliminate various threats to society but ensures that each cop's descendants, for millions of unborn generations, will also have that knowledge!

There can be no natural explanation for the origin of this cancer-fighting wonder other than past heavy losses of juveniles to cancer.*

Because it was activated only after healthy cells were converted into the deadly cancer state, the increasingly efficient immune system enabled many species to weaken, or even abandon, first line defenses. The animals were still, from the gene's view, disposable vehicles, and every act of somatic cell creation in a developing animal was still a threat to the germ line. But the "fail safe" nature of immune systems liberated the gene pools. Released from the restrictions imposed by risk-aversive cancer defenses, many of these emboldened invaders of the sun-drenched land surfaces could do what would be unthinkable with only a single line of defenses:

Increase the length of prereproductive life.

Lengthen total life spans. The aging process was attenuated.

Invest more cells in each organism. Giant animals--dinosaurs at an earlier time, humans now--came to dominate life on earth.

Externalize soft tissue as the need for noncellular external hard coverings were reduced or eliminated.**

---

*The immune system also fights viruses, bacteria and other threats to the germ line; many juveniles were killed by those pathogens. But these facts have no effect on the validity of my assertions about cancer selection's evolutionary role.

**Perceptual errors mislead us. Biologists usually refer to this phenomenon as the internalization of the skeleton. It is more enlightening, however, now that we know about oncogenes, to view this particular transformation as the emergence of soft tissue (made up of dividing cells) from behind protective coverings.

Because of that externalization of tissue, develop greater flexibility and mobility.

Eliminate, in some species, body hair, a noncellular covering with proven cancer-defense properties. (I say more about hair later in the chapter.)

Reduce skin pigmentation in many humans and in a few species of domestic animals--some pigs and some rabbits have white-pink skin.*

In many species, spend entire days in direct sunlight.

In most immunologically-equipped lineages the animals increased in size. That is because immune systems not only *permitted* larger animals, they *encouraged* them. With an effective immune defense in place additional cells actually protect against cancer.

But if *fewer* cells were cancer defensive in insects, how could *more* cells be cancer defensive in vertebrates? To understand this apparent paradox, consider two vertebrates with cells of similar size. One is a mouse whose liver is no larger than the eraser at the end of a pencil. The other is a whale, and it's liver is the size of a small automobile. If cancer were to start in one liver cell in each animal and proliferate at the same rate of speed, which animal would be the first to die? Obviously, the mouse would go first. Because of its smaller size, the mouse's liver would stop function-

---

*Even the reduction of pigmentation may have been caused by cancer selection. Recent research suggests that human breast cancer among Caucasian women is lower in sunny locations than in darker places. The researchers' hypothesis: Vitamin D, which is more easily absorbed by light skin, acts as a cancer defense.

ing before the whale's.* And the whale's immune system, with more time to organize a counterattack against the killer cells, would have a better chance of winning its fight against the killer cells and might save the animal.

(In his "Phylogeny and Oncogeny" Clyde J. Dawe pointed out that although whales have many more cells at risk than mice and might be expected to have higher lethal cancer rates they in fact have far lower rates. He speculated that certain physical characteristics of whales [he mentions higher levels of fatty tissue] might explain the whale's lower death rate. He seems not to have considered time-to-kill versus time-to-react as a factor.)

The terrestrial vertebrates include among their number the only *large* animals that regularly expose themselves to intense sunlight. Vertebrates are also the only animals known to have cancer-specific immune systems. And they have yet another unique characteristic: they are the only animals that sleep.

Sleep is a major evolutionary mystery.** Land vertebrates spend one-third of their lives in an unconscious state, utterly defenseless against attack by predators. Natural selection would have worked against the selection of this defenseless state unless it offered other life-or-death benefits. My theory looks at all the facts and asserts that sleep's primary function is to defend against cancer.

To begin my case, consider that the greatest risk of cancer initiation occurs during mitosis. That delicate process of passing genetic material from one mother cell to two daughter cells is, in organisms with oncogenes, nothing less than death-defying. It is

---

*Their small size makes mice attractive laboratory animals (easy to handle, cheap to maintain, etc.) but it may also explain their seemingly higher sensitivity to cancer. The "carcinogen of the month" phenomenon would probably disappear if researchers worked with larger animals.

**This is another problem that professional evolutionists seem to duck. Although the authors of some books about sleep speculate about its possible evolutionary function, I have yet to see it mentioned in an evolution text.

also an incredibly frequent occurrence in large animals. Cells divide ten quadrillion times during a human lifetime. *That's 350 thousand million cell divisions every twenty-four hours!* If just *one* of those divisions went awry, the mishap could kill the organism. And any cell divisions that misfired in juveniles would imperil the lineage.

Significantly, these highly dangerous acts occur in vertebrates *during sleep.* Human skin cells, for example, divide mostly between the hours of midnight and 4 AM. The connection with sunlight is obvious. Cells divide at night in animals that are active during daylight and during the day in most nocturnal animals. Bats and mice sleep during the day, but they sleep, and their cells divide (its been observed and measured in mice) in places sheltered from sunlight; bats sleep in caves and mice in burrows.

Using the "cause of death" rule, the universality of sleep in land vertebrates (all mammals, birds and reptiles sleep) leads to the question, what killed animals that did *not* sleep? The facts--mitosis during sleep, sunlight avoidance while sleeping--point to cancer.

Another set of facts that supports this idea is the age-related sleep pattern in our own species. Humans sleep most during infancy--newborns sleep 18 or more hours a day--when new cell production, and the risk of cancer initiation, is at its highest level. After infancy sleep decreases steadily with age, but with one significant exception. Adolescents sleep more than pre-adolescents. Again, there is a correlation with growth and increased cell division: rates of increase in height and weight during adolescence are second only to infancy. Cancer experience also correlates. Adolescents are especially vulnerable to cancer related to growth. Leg bones grow rapidly during adolescence and cancer in those bones almost exclusively occurs in teenagers.

Another medical fact pointing toward sleep's function as a cancer defense: the increase in sleep following severe trauma. Persons recovering from major surgery or other trauma--when cells division increases to repair damaged tissue--sleep more than normal.

There is still more evidence. The pituitary gland secretes

growth hormone when we sleep. According to Yasuro Takahashi, "...the highest peak of [growth hormone] concentrations in a 24-hour period always occurs during...sleep."

How does sleep enhance cancer-free cell division? I don't know. This is a black box proposal. I suspect, however, that the state of unconsciousness was selected to enforce physical inactivity and that inactivity provides an internal somatic environment conducive to the successful division of cells.

I have said that insects shield their larvae from solar radiation as a cancer defense. The terrestrial vertebrates also protected their embryos from radiation, but they didn't put them under rocks.

Most vertebrate fish embryos were not protected by their parents. They reproduced with external fertilization and external gestation; many fertilized eggs develop in open water. But when some of the fishes' descendants migrated to land they moved toward greater embryo protection. This is evident in the earliest land animals, the amphibians. Although some amphibians use the fish system of external fertilization and external gestation, others use internal fertilization followed by external gestation. And a few species use both internal fertilization and internal gestation.

In the next big evolutionary step, the emergence of true terrestrials, the reptiles and birds, fertilization became internal and all embryos were protected in hard-shelled eggs, some of which were buried by the parents.

Embryo protection was further intensified in mammals. Both fertilization and gestation are internal.

That progression from exposed fertilization and exposed gestation to shielded fertilization and shielded gestation implies unrelenting selection pressure. Such long term trends in many lineages are best explained, again applying Occam's Razor, by a single selection mechanism working throughout the long transformation period, rather than by a melange of assumptions.

Increased protection of embryos occurred in lineages that underwent great transformation and my theory says transformation itself could not occur without lots of juvenile cancer, including embryo cancer. The intensification of cancer selection pressure as the animals moved away from the protection of the sea would also explain the change to internal fertilization and internal gestation. As my theory would predict, no comparable intensification of protection of very young offspring occurred when *plants* moved from marine to terrestrial habitats.

Despite the fact that many mammals have discarded heavy external protection against sunlight, all land vertebrates continue to shield mitotic cells from natural radiation.

Blood cells in humans and other vertebrates, which divide more rapidly than other cells, divide inside large bones. As X-ray images demonstrate, bone tissue protects against radiation.

In four-legged animals, the soft organs, which are made up mainly of dividing cells, are protected from exposure to sunlight by layers of cells that do not divide; muscles and, to a lesser extent, nerve cells.

The observation that pre-mitotic cells are routinely shielded by cells that do not divide suggests that cancer selection explains one of the great mysteries of recent evolution, the origin of the human brain.

Paleontologists have established, with the 1974 discovery in Ethiopia of the hominid fossil "Lucy," that our ancestors first became bipedal about 3.5 million years ago. They stood up before they acquired their large brains. The big brains--they more than doubled in size from Lucy's--did not appear until about 2 million years ago. That sudden appearance--and in evolution 1.5 million years is a short time--is a puzzle. So quickly did the new brain appear that biologist Anthony Smith estimates that it grew at an average rate of 90,000 cells in each generation!

All previous ideas about that sudden origin revolve around the supposed survival value of human intelligence. They ignore several powerful signs pointing to cancer selection.

The *locale* where our ancestors were living when the big brains first appeared is highly significant. It was in the Rift Valley, which runs from North to South, dividing central Africa in half. West of the valley the land is covered with heavy foliage; it's mostly deep, dark jungle. To the east it's savanna; open land bombarded by fierce tropical sunlight. The valley itself, where Lucy lived, is now one of the hottest places on earth. It is risky to assume that current climatic conditions obtained millions of years in the past, but I make no such assumption. According to a 1984 article in *The New York Times*, specialists are convinced that humans appeared when the area changed from shady forest to sun-drenched savanna.

Suddenly spending entire days with the blazing African sun beating down on the top of their heads (thanks to their recent adaptation of bipedalism), early humans suffered losses from brain cancer. But--and this is essential--most brain tumors do not start in functioning brain cells, not in neurons. They start in glial cells, dividing non-nerve cells that circulate inside the cranial vault. Neurons are postmitotic; they never divide, not once the brain has been constructed. And brain construction is completed in *early* childhood.

If glial-cell cancer killed many human children, selection would have favored the placement of additional neurons on the top of the mammalian brain we inherited from Lucy and our other protohuman ancestors. Those additional nondividing cells, placed between the dividing cells and that harsh African sun, would have blocked the carcinogenic solar radiation.

Certain observations support this idea:

Cancer is the second leading cause of death among American children. And the *second* leading site of lethal cancer in children is brain cancer; it accounted for 14% of childhood cancer deaths in a recent year. (The leading cause of death is accidents and leukemia is the most common cancer.)

Children have thick hair *only* on the top of their heads. Humans lost their thick *body* hair, and biologists are convinced that they shed it to survive in the heat of the African plain.

But most of our body heat escapes through our heads. (It's why most people wear hats in cold weather.) If we got rid of body hair to keep cool in the African heat, its retention by juveniles (remember, their welfare was essential to lineage survival) in the one place where it would most interfere with body-cooling suggests that something else was also involved. I think childrens' hair protected them against sunlight-caused brain cancer. Hair's ability to defend against cancer has been established in experiments. Nude mice (they're shaved daily) exposed to ultraviolet radiation displayed increased tumor formation.

Our big brain's intellectual capacity is far in excess of what was needed to survive. Even Alfred Wallace, its codiscoverer, was convinced that natural selection could not explain the human brain. He argued that natural selection would produce the sufficient, but not the supererogatory. Wallace was right.* There is no conceivable survival purpose for a brain capable of knowing how to play bridge, write symphonies or create new theories of evolution.

But our genes nonetheless did amass a lot of additional non-dividing neurons on top of Lucy's mammalian brain. Using the "cause of death" rule, we must ask how did those creatures without the new mass of brain tissue actually die? What killed them?

There is, of course, only one answer consistent with all the facts: they died of brain cancer starting in those non-neural glial cells inside the cranium.

Although the increased intelligence provided by the enlarged human brain undoubtedly helped our ancestors to survive, any *successful* theory of the human brain's origin should explain why it appeared when it did--soon after hominids stood up--and what killed young animals who were not equipped with the new improved

---

*He was right, that is, about the inadequacy of the natural selection explanation. He was wrong to attribute our brain's origin to supernatural causes.

organ.*

The idea that cancer selection caused the origin of the human brain may strike some as overly audacious. But it is based on the only idea in this book--that non-dividing cells protect dividing cells from cancer--that can be tested experimentally. I have designed a simple experiment for that purpose and describe it in Chapter Eleven.

The terrestrials are the most transformed of all animals because their lineages endured more cancer selection than any others. The engine of transformation ran faster and produced greatly changed animals.

One way to demonstrate how transformational evolution accelerated in the harshly carcinogenic terrestrial habitat is to compare a large terrestrial mammal to an animal that never left the marine environment.

At one of the two scientific meetings I have attended, a renowned biologist remarked, in the midst of a wide-ranging lecture, and for reasons best known to him, that cows were no more complex than sharks. That professional scientist did not, I am sure, actually mean to select a specimen of domesticated cattle, or--adding sin to error--to single out the female of one species to make his point. Domesticated animals are off-limits to those of us active in evolution for the obvious reason that they were artificially bred to have characteristics that made it easier for humans to dominate them. A comparison of cows with sharks is particularly unhelpful because cows were bred to be both docile and stupid. It

---

*The selection of a character for one function (cancer protection) and its subsequent use for another beneficial purpose (problem solving) is called preadaptation. It is an established element of neo-Darwinism.

It is also possible that bipedalism caused other increases in cancer. I am not aware of any studies comparing the incidence of human breast cancer or cancers originating in the abdomen with comparable data for larger tetrapods. The latter animals afford (by virtue of their horizontal configuration) their mammary glands and soft internal organs more protection from sunlight than do humans and may, as a result, incur less cancer at those sites.

would be fair, however, to compare a *wild* bovine to a shark to see if the big land animal is any more complex than the big marine animal. I will do just that, using the African Cape Buffalo as my example.

According to Dorst and Dandelot's *A Field Guide to the Larger Mammals of Africa*, these massive bovines (they can reach more than five feet in height at the shoulder), are known to move in massive herds of up to 2000 individuals. The great mass of animals is dominated by a master bull and (no sexists, these beasts) a senior cow. Those co-generals of what amounts to an army of buffalo depend on scouts--other buffalo that fan out from the main body of the herd--to warn them of approaching danger. This is truly remarkable behavior. Could our earliest human ancestors have done any better when moving through hostile territory?

In addition to that extraordinary organized group behavior, which suggests a level of intelligence far above that of sharks, the African Cape Buffalo has a reputation for savagery. Not only do hunters consider them the most dangerous of all African game, these beasts, unlike most African grazing animals, do not flee from approaching lions. Instead, when the big cats approach the herd the adults gather around the young calves and face the lions, fully prepared to gore any attacker. The lions usually slink away. Of course, diseased buffalo and even youngsters occasionally succumb to predators, but unlike old Bossy in the barn, these wild bovines fight ferociously when attacked.

Sharks are also ferocious, but their savagery is quite another matter. They're carnivores and their daily survival depends completely on their ability to violently overwhelm and then kill other marine animals in order to eat them. Shark violence is simply a genetically-determined means of obtaining food. Cape Buffalos, however, are herbivores; they don't kill to eat. And because their ability to fight when attacked is not related to the need for nourishment, its evolution involved additional transformation. In their lineage two separate complex systems evolved, one to gather food in a peaceful manner and another to wreak violence on would-be predators. Sharks needed only one system. (Although there are fish who graze and fish who attack other fish, I know of no fish species that survives on plant matter and whose members routinely kill big predators.)

The shark is nonetheless a complex animal and, as I have been saying, the old theory of evolution cannot explain *any* complex animals including sharks. But theories of evolution are supposed to explain *transformation*--the creation of complexity--as well as complexity itself and in comparing the history of sharks with bovines it is clear that there was a lot more transformation in bovine lineages than in the sharks'.

The shark phylum is very old. According to the fossil record the earliest sharks appeared about 350 million years ago. Since that time sharks have evolved somewhat, having produced about 200 species of sharks and 300 species of rays. But sharks and rays are still sharks and rays. Sharks may be fascinating (and terrifying) predators, but that's all their lineages have been able to produce. For that reason most biologists consider them evolutionary dead ends.

But if we examine the lineage of the African Cape Buffalo over the same period we will find a profoundly different history, for when the first sharks came into existence they may have encountered the buffalos' ancestors. They were vertebrate fish.

So great was the transformation in the fish-to-buffalo lineage that it would be impossible to list all the physiological changes that occurred, but it is apparent that it first required the emergence of amphibian capability, then reptilian characteristics, and still later, mammalian characteristics. There was a switch in reproduction from the fertilization of eggs in the open sea to internal fertilization and then to internal gestation, the change from cold-bloodedness to warm-bloodedness, a drastic change in digestive systems in order to survive on rough plant matter--etcetera, etcetera, etcetera. All those gross transformations required radical changes in development programs and processes and in the constitution of individual cells.

It ought to be clear by now that it was foolish of that famous biologist to utter his strange comparison of the complexity (which is to say, the evolutionary history) of sharks to that of bovines, or,

for that matter, to any of the numerous land descendants of vertebrate fish.

The evolutionary changes needed to produce land vertebrates
simply could not have occurred without greater losses to cancer
selection than those endured by the virtually unchanged shark
lineages. And since my theory says that the operation of the engine of transformation meant that increases in complexity caused
increases in cancer selection, it should come as no surprise that its
corollary--less change in the animals meant less cancer selection in
the lineage--is borne out by modern sharks. Genes that have been
producing nothing but sharks (and shark-like rays) for 350 million
years have the learning curve of self-manufacture well behind them.
That is the evolutionary reason why modern sharks have little
cancer; their cancer defenses have not been challenged by changes
in development programs to the extent experienced by terrestrial
vertebrates.

Theories whose predictive ability holds up under extreme conditions
are superior to those that falter.

If sunlight-initiated cancer selection transformed simple animals
to complex animals what happened if lineages were *totally* deprived
of sunlight for many millions of years? There are such lineages
and the animals they produced support my theory--in their own
peculiar way.

No terrestrial animals avoid sunlight more thoroughly than
those that live *inside* other organisms. Internal parasites like the
common tape worm haven't seen sunlight for several hundred million years. And, of great theoretical significance, these parasites
have done something evolutionarily extraordinary. They have *lost*
complexity! One authority [Huff] says tape worms "lost practically
all trace of free-living characteristics," that they have no digestive
system and resemble "a colonial form in consisting of many, fairly
autonomous parts."

These extreme sunlight-avoiders resemble colonial organisms
in another way. In at least certain species they are capable of

exceptional longevity; tape worms that inhabit human guts can live for 35 years. Large simple organisms that live for a long time--trees, sponges, certain cnidarians--are prevalent in nonanimal lineages. But simplicity and longevity in the same animal is unknown--except for these internal parasites.

Again, cancer selection explains. Cancer defenses enabled DNA to create complex organisms but in tapeworm lineages, prolonged relief from the pressure of cancer selection (by depending on the defenses outside themselves, in the body of the host animal) weakened the internal cancer defenses and lowered the tape worm DNA's capacity for precise development. The worms regressed to tissue-level simplicity.

By regressing tape worms indirectly confirm my central idea that exposure to cancer-causing sunlight caused complexity.

Some readers may feel overwhelmed by my insistence that so many separate characters and traits of land animals were cancer-related in origin. They may think I go too far. But the most spectacular accomplishment in those lineages, especially those of insects and vertebrates, was not the accumulation of this feature or that character. It was the sheer breath-taking magnitude of transformational evolution itself. From the time they left the sea, surviving terrestrial animal gene pools--spectacularly efficient directors of animal-manufacturing systems--added an immense number of somatic revisions to *already* complex organisms, and they did it without once breaking the chain of successful replication in breeders. Anyone who thinks that could possibly have happened without intense selection pressure from cancer should reflect upon the arguments made in earlier chapters. Rather than overdoing it, I have probably missed many manifestations of that pressure.

Ten

A Final Look at Nonanimals

What Shakespeare said of old age can well be applied to the
coelenterates: "Sans teeth, sans eyes, sans taste, sans everything."

Richard Headstrom

So what happened to the nonanimals? What did natural selection accomplish during those hundreds of millions of years while
cancer selection was transforming worms into elephants, giant squid,
social insects and human beings? As my theory would predict, not
much. *None* of the products of evolution by means of natural
selection are as complex as *any* of the products of evolution by
means of cancer selection *and* natural selection.

And just what is complexity? Although biologists seem unable
to agree on any single measure, various authors have cited the
number of *kinds* of cells in the organism as a good indicator. They
are right. It is several orders of magnitude more difficult to start
with a zygote and construct a trillion-cell organism containing 200
different kinds of cells--a human being--than it is to construct a trillion-cell organism having about ten types--a giant jellyfish. Cell
differentiation is an excellent "rule-of-thumb" indicator of relative
complexity.

Here's a summary adapted from John Tyler Bonner's *The Evo-*

lion-cell organism having about ten types--a giant jellyfish. Cell differentiation is an excellent "rule-of-thumb" indicator of relative complexity.

Here's a summary adapted from John Tyler Bonner's *The Evolution of Complexity by Means of Natural Selection*:*

Number of Cell Types

Nonanimals:

Mushrooms and kelp	07
Sponges, cnidarians	~11
Giant sequoia	~30

Animals:

Fairy flies, squids	~55
Fish, frogs, whales	~120

This comparison is consistent with the assertion that animals are more complex than nonanimals.

I said earlier that in evaluating evolutionary mechanisms, the complexity of sexual organs must be distinguished from complexity in tissue not directly in sexual reproduction. I will now explain why.

Natural selection worked *directly* on reproductive systems. In all sexually reproducing lineages--nonanimal as well as animal--selection pressure focused most intensely on those cells, tissues and organs directly involved in reproduction. The reason for this is clear. Organisms without precisely-made and efficiently-operating sex cells and sex organs would not have bred. Their genes would have died. That evolutionary logic is in accord with the reality of what happened; in plants the most complex parts produced by natural selection--one might say their *only* complex parts--are their

---

*Alas, despite the title of his book, Professor Bonner does not explain how natural selection caused animal complexity. His book is largely descriptive.

reproductive organs.

The complexity of plants' sex organs was based on the need for extreme *uniformity*. Sameness of sex organs was essential. It's difficult to even imagine a species burdened with an assortment of different types of sexual equipment. Without extreme organ-uniformity throughout the population, sexual compatibility would be a hit-or-miss matter. We can be sure of it: no such species now exists or ever existed.

In constructing those hyperuniform reproductive systems nature required no feedback devices inside every somatic cell (functioning oncogenes) to inform gene pools whether or not the cells in the systems had been manufactured properly. If the cells and the sex organs they comprised did not function efficiently, feedback was accomplished with brutal efficiency: the DNA of the organism was eliminated from the gene pool.

But consider the entirely different situation of cells *not* directly involved in sexual reproduction. Consider, for example, plant cells in root or branch tissue. Those *individual* cells never played a crucial role in survival of the lineage. If one cell misfunctioned because of a gross somatic mutation, many other cells, assigned to the same task, would carry on. Because they lacked the functioning oncogenes that would have given them the ability to kill the organism, individual plant cells could never communicate with gene pools. Neither could cells in the tentacles of a Portuguese-Man-of-War. Or those inside the body cavity of a sponge.

With the sole exception of reproductive organs, selection in nonanimals could "work" only on the entire ensemble of cells--on whole organisms--not on the molecules inside the nuclei of cells. If the body survived and reproduced, then the imperative of germ line survival was met. Nothing further was demanded and nothing further was wrought. Without cancer selection there was no gene-determined mandate for precision and uniformity throughout the organism. And without precision and uniformity, complexity--*animal-like complexity*--could never evolve.

But sex organs are another matter. That's where evolutionary logic tells us to look for complexity in plants and that's where botanists find it. And some of that complexity can be fascinating. Recently, two botanists, A. Dafni and P. Bernhardt described degrees of complexity in certain orchids that are truly astonishing. They discovered one species that creates objects inside the flower that look like rotting fruit. Fruit-eating insects enter the flower to investigate. Once they discover they've been tricked the bugs fly away, but with a load of the orchid's pollen attached to their bodies; the lure works to the plant's advantage. Other orchids construct "dummies" of female wasps inside their flowers. Male wasps attempt to copulate with the phony females. As with their friends who are tricked by the fake food, these would-be lovers are bamboozled into serving the orchid's interests; they too fly off loaded with pollen.

Those examples enchant us (and probably make life less boring for botanists) but they occur only where my theory says they could, as part of functioning sex organs.*

Certain fundamental genetic differences between plants and animals are unexplained by the old theory of evolution. Plants, according to Ernst Mayr, practice "widespread clandestine hybridization", successful breeding in the wild by individuals of two different species. Oxford biologist Mark Ridley says hybridization may account for the origin of as many as one-third of all plant species. But hybridization is virtually unknown in animals.

According to Mayr, polyploidy, the spontaneous multiplication of chromosomes during sexual reproduction, occurs in plants but is virtually nonexistent in animals. Mayr's explanation for these profound differences--"Higher animals simply are [sic] different genetic systems from plants."--strikes me as nothing more than a shrug, suggesting either that Professor Mayr doesn't think those dif-

---

*Of course the origin of orchids and other flowers coincided with the emergence of insects, birds and other pollen-transferring animals. And since those animals' evolution resulted from cancer selection, plants with flowers would not have evolved without it.

ferences warrant further theoretical attention, or that he hasn't a clue about their origin. Indeed, in keeping with the well-established practice of keeping mum whenever they encounter inexplicable phenomena, neo-Darwinians are generally silent on these matters. Another important difference between animals and nonanimals involves the sequestration of sex cells. Animals sequester them and plants don't. Because their gametes are direct offspring of ordinary somatic cells, it is possible that somatic mutations in plants can be passed to gametes and become evolutionarily significant germ-line mutations. But because of early sequestration of animals' gamete-producing cells, animals' somatic mutations can not affect the genetic makeup of future generations. (Mutations that originate in cells ancestral to gametes are, in my definition, germ-line mutations.)

These fundamental differences between plant genetic systems and those of the animals--hybridization, polyploidy and non-sequestration--deserve more than our idle interest because, unfortunately for neo-Darwinists, they clash with another significant fact.

If plants commonly hybridize in the wild, frequently double or triple their chromosomes during reproduction, and pass some somatic mutations to future generations, logic (and the old theory of evolution) would tell us to expect genetic diversification, as measured by speciation rates, to be higher among plants than in animals. After all, hybrids differ genetically from both parents to an extent simply not possible in an *intra*species union. And if chromosomes double or triple spontaneously, and if ordinary somatic mutations can lead to gamete mutations, then the chances of creating, each time a zygote is formed, something that is genetically completely different from its parents are substantially increased. Those powerful genetic "wild cards" ought to have created many potential founders of new species; plants would seem to have had open to them avenues to speciation and diversification that were closed to animals.

But when it comes to actual speciation, nature ignored the "advantages" offered by those genetic devices. Animals species greatly outnumber plants. Animals species total more than five million plants, about half a million. What happened? Why didn't the plants speciate--and diversify--more than animals?*

My theory solves the problem. It asserts that differences in the DNA's ability to create variety in zygotes--which is to say molecular mechanisms recognized by geneticists--were *not* a major factor in the origin of species. What *did* count was the relative power of the DNA to convert zygotes to organisms. Complete mastery of development--extremely high program implementation-fidelity--was essential to speciation. Thanks to cancer selection, animal DNA acquired the power to create organisms with highly *specific* characteristics anywhere in their bodies. Greater control over organism construction permitted more organismic variation--genetically-determined variation--and greater speciation potential.

Plants never acquired that power. Without it, their access to those seemingly more productive genetic mechanisms was of little use.

The role of symmetry in nonanimals is worth examining.

A general radial symmetry of whole organisms can be seen in trees and in certain cnidarians, including the common disk-shaped jelly fish, and in siphonophores like the Portuguese-Man-of-War. Some biologists even perceive bilateral symmetry in the siphono-phores. However, if it exists, the bilateral symmetry found in those cell colonies would seem to be the result of natural selection working on mature organisms. Unlike true animals, the symmetry appears in adult forms, not in the earliest stages of development. I've mentioned that in echinoderms, an animal phyla that includes the starfish, just the opposite happens; lateral symmetry, pronounced in the larvae, disappears as the animal matures.

Another interesting point about symmetry in multicells is that

---

*Some may think animals' mobility better equipped them to establish offshoot populations (potentially new species) in distant habitats. But the fact that individual plants cannot move has not impaired their geographic diversity. Pollen, seeds and even whole fruit, e.g., coconuts, are transported hundreds or even thousands of miles by wind, water or cooperative animals.

in animals strict bilateral symmetry is found only in the whole organism, or in those organs bisected by the longitudinal axis of the body. Just consider the human body. Yes, our bodies are symmetrical, and so (more or less) are our skulls, brains, noses, and sex organs. But there is nothing symmetrical about the individual ear, lung, hand or foot. The symmetry is in the whole and not, except for those organs bisected by our long axis, in the parts.

But if we look at nonanimals we see an entirely different type of symmetry. A gnarled, centuries-old oak tree growing atop a wind-swept hill would, because of the effect of the wind, display little symmetry in the arrangement of its branches (or, if we could see them, its roots) but each of its leaves would be bilaterally symmetrical.

So we have two very different kinds of symmetry to explain. In animals, symmetry of the whole organism (during the lifetime of most, in the larval stage of all) but not in individual parts. In plants, little symmetry in the whole organism, but bilateral symmetry in some individual parts. What do those facts (ignored by old-theory evolutionists) tell us?

They imply that the symmetry of oak leaves was selected--by *natural* selection--because it was suitable to the function of the leaf itself. For one thing, symmetrical leaves would be less likely to be blown off in high winds than those with asymmetrical shapes. For another, the flow of the products of photosynthesis from leaf tissue to the rest of the tree was facilitated by a symmetrical network of vessels. Trees with more efficient leaves were more likely to survive.

As I point out in Chapter Seven animal symmetry was selected long before any of its functional advantages (in the ability of birds to fly, for example) could have been utilized. Its appearance in the larval stages of all animals, even in those that do not exhibit it in adult forms, strongly supports the idea that it provided a survival benefit to the first animals and that it did so at the time in the organisms' life when it appears in modern organisms, during the earliest high-growth stage of embryogenesis.

Cancer avoidance in animal lineages and zero cancer selection

in nonanimal lineages is the only logical explanation for the profoundly different relationship with symmetry in animals versus the
other multicells.

Throughout this book I hammer away at the lack of vital organs
in plants and other nonanimals. I am certain that no such organs
are found outside the animal kingdom. But in order to be absolutely sure that I am not overlooking any wondrously complicated
parts and devices I want to briefly consider what are perhaps the
most complex of the nonanimals: carnivorous plants and siphonophores.

Perhaps the closest thing to a vital organ in a nonanimal is
found in carnivorous plants. Those plants, which devour insects
and, in a few cases, small frogs and other amphibians may represent the ultimate achievement of evolution without the aid of cancer
selection. Growing in regions where soil is poor, those plants
adapted by consuming animal protein for essential nutrients. The
mechanisms for enticing the insects, entrapping them and digesting
them would seem to represent one of the peaks of somatic complexity achieved by natural selection. (It is worth noting that the
insect-entrapping tissue is sited in the flower, which, as a sex organ,
is where we ought to expect plant complexity.)

Another candidate for the champion of complexity created by
old theory mechanisms are siphonophores like the Portuguese-Man-
of-War. That organism (or is it a cell colony?) survives on fish
that accidentally swim too close to its transparent tentacles. It kills
the fish with nematocysts or stinging cells and digests it within its
central cavity. Some siphonophores even have eye spots, used, perhaps, to sense the presence of prey. However, they have no brain-
-only a primitive network of nerve cells--so it is unlikely that those
organisms' can "see" with anything near the acuity enjoyed, for
example, by the common house fly. With their stereoscopic vision
and their wonderfully efficient tiny brains (which process the data
gathered by the eyes), flies are capable of remarkable feats of precise, sight-controlled coordination; they can alight on a string

dangling from a light fixture. Portuguese-Men-Of-War are incapable of anything more than weak control over their locomotion and are propelled, hither and yon, by the wind pressure on their trademark gas sacs.

As this brief examination shows even the complexity of carnivorous plants and siphonophores doesn't come close to that of *any* animal formed by rigorous cancer selection.

One legitimate question that might be asked about the evolution of plants and the other nonanimals is how did those lineages cope with the mutagenicity of sunlight? After all, even if they did not get cancer or have complex organs, the potential for that most pervasive of mutagens to wreak havoc with development would have been a threat to even simple multicells. Also, because, plants, unlike animals, did not sequester somatic cells ancestral to gametes, sunlight-induced somatic mutations could lead to excessive germline mutations and threaten the survival of the lineage.

The solution of the ancient plant lineages to the deleterious effects of sunlight can be deduced from one fundamental characteristic of modern plants and other simple multicells: abundant somatic redundancy. Animals tend not to have a surplus of vital parts, but trees have many branches and roots all performing the same functions. The redundancy of sex organs--some trees produce thousands of flowers--is obvious. Redundancy significantly lessened the damage to the germ line that could be caused by solar radiation. A "hit" of radiation on a primordial tree might have caused a deadly mutation in one branch, but with no possibility of dying of cancer and with many alternative branches, the tree's survival and that of its lineage would not have been jeopardized by the event. Similarly, if a mutation damaged a gamete in one flower the abundance of other identical blossoms ensured that many undamaged gametes were available.

Nature conducted a long and comprehensive test of the tranformational power of the two types of selection. In one large group--plants, cnidarians and sponges--the results show that with natural selection as the sole transforming mechanism the most complex forms are less complex than even the simplest of the animals. In the other large group, which benefitted from both cancer selection *and* natural selection, the results are the most complex of all known objects. During the many hundreds of millions of years that natural selection needed to come up with Venus Fly Traps and gas-bags-that-catch-fish as its outstanding accomplishments, the lineages created by cancer selection were producing, with aplomb, a great variety of stupendously complex organisms in millions of different lineages.

That great natural experiment, whose results are there for all to see, tested the validity of the two competing theories of evolution. The old theory is shown to be sufficient to the task of explaining the evolution of plants, jellyfish and sponges. We can all agree with the biologists: natural selection did it without cancer selection. However, it is far beyond the power of the theory of evolution by means of natural selection to explain--to cite just one example--the common house fly.

Part II

Evaluations

Eleven

Modern Cancer Evidence

Sit down before fact as a little child, be prepared to give up every preconceived notion, follow humbly wherever and to whatever abysses nature leads, or you shall learn nothing.

Thomas H. Huxley

Cancer facts support my theory.

Reaching that conclusion required a great deal of work. The complexity of cancer selection's theoretical role in evolution and the mass of information available about cancer-as-a-disease demanded careful analysis. First, I had to separate the relevant facts from the irrelevant. Then I had to analyze the selected facts from an evolutionary view point (completely disagreeing, in one important instance, with a well-known cancer authority) and then I had to test my analysis for conformation with the theory.

The conclusions I reached at the end of that effort are organized under headings that refer to different parts of the theory.

The initiation of cancer by a mutational event. My entire theory rests on the proposal that cancer begins with a replication error--a mutational event--in a single somatic cell. If I am correct, then cancer selection's potential for great evolutionary influence is established: it functioned as a *de facto* enforcer of fidelity during development.

Although Bruce Ames and his colleagues published their discovery of the correlation between carcinogens and mutagens in 1975, the possibility that cancer starts with a somatic mutation was suspected much earlier. Its original proponent was Theodor Boveri, a nineteenth-century German scientist who noticed that cancer cells often contained damaged (mutated) chromosomes. That led him to propose that cancer began with chromosome damage. According to Errol C. Friedberg (in *Cancer Biology*), a modernized version of Boveri's hypothesis might go something like this:

> Carcinogens are agents that interact with and cause damage to DNA. If not properly repaired, the damage becomes fixed in replicated DNA as mutations. Mutations affecting one or more genes involved in growth regulation can result in neoplastic transformation.

Eventually, Boveri's idea ran into competition, research that seemed to contradict it. In 1911 Peyton Rous, a researcher at the Rockefeller Institute, transplanted chicken sarcomas by injecting previously healthy birds with a fluid drawn from cancerous cells. The fluid had been passed through filters too fine for cells. That left viruses, which are too small to be trapped by filters, as the likely transmitters of the disease. Although his discovery met with widespread skepticism at the time, thirty years later investigators verified his finding. Viruses do indeed cause what is now called "Rous's sarcoma". By then in his eighties, Rous was awarded a Nobel Prize following the belated verification.

Howard M. Temin, an American researcher, reconciled the two conflicting models of cancer-initiation. He and his colleagues showed that viruses that initiate Rous sarcomas in chickens (and in some other animals, but not in humans) do so by causing changes in the cellular DNA of the host animal. The virus invades the cell nucleus and interferes with normal replication. It causes cancer because it is *mutagenic*. Temin's conclusions are consistent with my premise that all cancers begin with mutational events.

The concept of mutation-causation is also consistent with cancers that occur following traumatic damage to tissue. Wounds increase somatic cell production--and the odds that one misstep in cell replication will cause lethal transformation. (I say more about these cancers later in the chapter.)

Certain human cancers are associated with *germ line* mutations. Children born with chromosome defects are more likely to develop cancer than normal children. Retinoblastoma, a form of eye cancer that afflicts children, is traceable to a specific congenital chromosomal defect. Children with other chromosomal defects that lead to noncancer symptoms, such as Fanconia's anemia and Down's syndrome, exhibit tumor-proneness measurably higher than normal individuals. In these cases, the DNA that was mutated in zygote formation was transferred, during development, to all somatic cells. Cancer begins in one of the somatic cells. These cases do not conflict with my theory. Even though the underlying cause of the cancer was an event that occurs before somatic cells were formed, (i.e., a germ-line mutation) it is the subsequent *expression* of the flaw in a somatic cell that initiates the lethal process.*

As mentioned earlier, molecular biologists have determined that only mutated oncogenes create cancerous cells.

In short, *all* modern scientific evidence supports the idea, central to my theory, that mutational events in somatic cells trigger cancer.

The existence of oncogenes. Molecular biologists discovered oncogenes in 1981. By now, according to a recent report in *The Economist*, more than 50 different types have been found in animals.

The existence of anti-oncogenes. Although no anti-oncogenes had been discovered at the time my theory was published in the *Journal of Theoretical Biology* they were found inside normal

---

*These cancers suggest the possibility that cancer selection may have encouraged accumulation of genes that reduced the incidence of germ line mutations. However, because most germ lines mutations are deleterious, natural selection would have favored measures that minimized them even without additional pressure from lethal cancer during development.

mammalian cells in 1985. Researchers noticed that some cancers start when certain genes are removed and concluded that the deleted genes function as anti-oncogenes. According to Raymond Ruddon in *Cancer Biology* "...[oncogenes] are under stringent control in human cells. The genes involved in this control [over oncogenes] have been termed 'antionco-genes'[sic]."

Of course, my theory considers many other genes, such as those for sun avoidance, to be anti-oncogenic in origin. The historical role of those genes as cancer-preventers would be difficult, if not impossible, to test in a laboratory; however, significant modern evidence for the cancer-defensive nature of characters formed by actions of those genes exists. I cover them later in the chapter.

The great age of oncogenes. Two of the scientists who discovered them assert that oncogenes must be at least as old as I claim. Ben-Zion Shilo and Robert A. Weinberg wrote in *Proceedings of National Academy of Sciences* in November 1981:

> We conclude that the common precursors of [oncogenes] were already evolved 800 million to a billion years ago...

Shilo and Weinberg are molecular biologists, not evolutionary theorists. They nonetheless used sound evolutionary logic in concluding that because oncogenes were found in animals as distantly related as insects and vertebrates, they must be of great evolutionary age. The alternative idea, that nature "invented" cancer-causing genes more than once, in two or more separate lineages must be judged, by rigorously applying Occam's Razor, inferior to singular origin.

Another modern finding consistent with the idea that oncogenes and cancer selection are of primordial origin is the sheer complexity of cancer defenses inside the cell. *The Economist* has reported that human oncogenes have been known to initiate cancer following *as many as ten* mutations. This suggests that defenses against initiation are old, powerful and complex. Complex defenses against *any* threat to the germ line are explained only by a long history of genetic deaths caused by that threat.

The role of oncogenes in development.   I refer briefly in
Chapter Two to a plausible function for oncogenes and aggressive
growth in embryogenesis, and I amplify that idea in Appendix I.
Modern research supports this view.

Perhaps the first evidence that embryonic growth and cancer
are related was uncovered in 1963 when Garri Israelivich Abelev,
a Russian scientist, observed that alpha-fetoprotein (AFP), a substance known to be produced early in fetal development, was also
present in the sera of mice with liver cancer.   According to two
American scientists, William Fishman and Stewart Sell, Abelev
"recognized the significance of his finding; that a fetal gene was
reexpressed in neoplasia."   Abelev's discovery led to a flurry of
investigations.   An American team found AFP in mouse livers that
had been injured and in which cells were being regenerated following the trauma.   Other researchers discovered that another substance, CEA, was in both fetal intestinal tissue and in colon
cancers.   As with AFP, CEA was not found after fetuses were
formed--except when cancer was present.   CEA is only one of a
family of glycoproteins produced by genes both in carcinogenesis
and during fetal development.   Other gene products found both in
fetal development and in carcinogenesis include fetal aldolase, fetal
glucose-ATP phosphotransferases, and a group of isoproteins.

There are other such substances, but listing them is not
essential.   What is important is that there are a lot of them, that
they are produced by genes, and that they are found only when cell
proliferation is *beneficial*--during the formation of a fetus or when
tissue is being regenerated following injury--or when it is *deadly*--
during the formation of malignant tumors.

These findings (all made before oncogenes were discovered by
molecular biologists) support the idea that genes initiate aggressive
growth in human fetuses and are normally suppressed as the animal
matures.   Because gene expression, i.e., its synthesis of specific
proteins, can only be turned off by other genes, these findings also
support my conclusion that anti-oncogenes must be present in every

cell.

The existence of adaptive pro-oncogenes. My theory claims that selection of genes for changes in phenotypes would frequently have been followed by increases in cancer selection. Transformational evolution kept cancer alive as a killer of juveniles.

Obviously, it is not possible to travel back in any lineage to see if new types of cancer came into existence concurrent with the introduction of major phenotypic changes. However, if this idea is correct, there ought to be clear indirect evidence in its support, especially in our own species. We humans are probably one of the more recently evolved of all vertebrates, and we are the one species for which a lot of cancer-related information is readily available.

But before looking at facts that support my concept of adaptive pro-oncogenes, I want to consider what sort of cancer experience we should expect if mine were an *invalid* theory. In other words, what cancer patterns would we have if the *old* theory of evolution were correct?

The answer to that question is simple. Without the introduction of genetic changes that themselves caused increases in developmental cancer, we would expect that nature would long ago have eliminated cancer *as a killer of children*. With some 800 million years of evolution behind it, natural selection would have overwhelmed the oncogenes' ability to destroy immature animals. Would there be cancer deaths in old adults? Possibly. Lineage survival is not affected by cancer in grandparents. But no child would die of it.

The facts are not in accord with what the old theory logically tells us to expect. In the case of human leukemia, the most common form of lethal cancer, there are two peaks of incidence. One occurs in persons more than 85 years old. No problem there for the old theory; that is what it tells us to expect. But the other peak occurs at a time in life that utterly destroys any pretense that the old theory is consistent with facts. *The other peak in leukemia incidence occurs among children ages zero to fourteen.*

My theory explains childhood leukemia through the mechanism of adaptive pro-oncogenes, genes for adaptive changes that were acquired at a price of increased juvenile cancer. The peak in childhood leukemia says that some human genotypes are not capable of creating humans that produce white blood cells with the level of exactness mandated by our oncogene-enforced genetic program. The process of eliminating those imperfect genotypes is continuing through the only conceivable selection agency, lethal childhood leukemia. Other genotypes are capable of carrying out the production of white blood cells during childhood and on into middle age, but their defenses against lethal replication errors (perhaps errors different from those that occur in children) do not function perfectly in old age. Those genotypes account for the second peak in leukemia incidence, the one that occurs in old age. Both peaks fit the theory of adaptive pro-oncogenes.

Further support the idea that increases in complexity were accompanied by intensified cancer selection comes from molecular research. Professor Fritz Anders of Justus-Liebig-Universitat in Giessen, Germany, and his colleagues summarized (in 1986) the correlation between complexity and the number of oncogenes in various animals. This is an extract of their findings:

Type of Oncogene in Normal Cells

Crabs	*ras*	*src*	*abl*				
Amphioxus	*ras*	*src*	*abl*	*sis*			
Sharks	*ras*	*src*	*abl*	*sis*	*yes*		
Frogs	*ras*	*src*	*abl*	*sis*	*yes*	*ros*	
Mammals	*ras*	*src*	*abl*	*sis*	*yes*	*ros*	*mos*

This pattern--selection of more cancer-causing genes in lineages that manufactured increasingly complex animals--strongly supports my claim that transformational changes in lineages caused concurrent intensification of cancer selection.

More support for the concept of adaptive pro-oncogenes comes from cancer-incidence data. In a 1989 paper Anders noted that "mammals are more afflicted with [cancer] than any other[s]..." Mammals include in their numbers a higher percentage of more recently transformed animals than do fish, for example, or reptiles.

Anders also cites higher cancer rates in artificially bred animals. Special strains and hybrids exhibit higher rates of spontaneous tumors than do natural populations, and investigators initiated tumors in the man-made strains more easily than in wild types. Among the tumor-susceptible, human-bred animals identified by Anders: special strains of fruit flies, domesticated trout, hybrid ducks, mice bred for laboratory use, Lipizzaner horses, domestic dogs (boxers seem to be especially vulnerable) and domestic cats. Because of documented human interference in their provenance, we *know* that those animals are recently evolved. Although their phenotypes do not differ greatly from their wild relatives (they have no new organs) minor man-made changes--which introduced new development processes--from the wild type appear to have created measurable increased opportunities for cancer initiation.

This is most significant. If *minor* genetic tampering by humans increased cancer incidence it is simply inconceivable that *major* evolutionary revisions in animal lineages--those needed to change worm-DNA to elephant-DNA--did not increase cancer selection pressure.

Other support for the existence of adaptive pro-oncogenes is found in the pattern of other childhood cancers.

It is apparent from fossils and historical records that we are taller than our ancestors. Although changes in nutrition may partially account for that increase, it is reasonable to presume from the facts that changes in the development program--the expression of genes for tallness--played a significant role. (Biology's well-known Cope's Law says that increased size was the norm in vertebrate evolution.)

According to my theory, those changes caused increases in childhood cancer in the affected tissues. Modern cancer data--specifically, bone cancer statistics--support me. According to Altman and Schwartz, (*Malignant Diseases of Infancy, Childhood and Adolescence*), osteosarcoma, a form of bone cancer rare in infancy, increases in incidence during the teenage years--when bone growth accelerates. Importantly, the authors note that "children with malignant bone tumors have been found to be *significantly taller than average*" [Emphasis added.] This suggests that genes for tallness carry a higher risk of cancer in cells directly involved in the increase in height.

Altman and Schwartz also report that dogs artificially bred for large sizes also incur higher rates of bone cancer.

> bone sarcoma in giant dogs, such as the St. Bernard and Great Dane occur nearly *200 times more frequently* than in small and medium sized breeds. [This] also suggests a link between greater bone growth and bone tumors. [Emphasis added.]

Evidence of high incidence of cancer in tissue that has undergone a known *recent* evolutionary increase in mass in *two* groups of animals (humans and canines) that have not had a common ancestor for scores of millions of years or longer is several orders of magnitude more convincing than evidence from a single species. These data strongly support the much-needed (by evolutionary biology) explanation for occurrence of juvenile cancer some 800 million years after the origin of oncogenes. They support the concept of genes that were selected because they were adaptive despite the increase in cancer rates following their selection: adaptive pro-oncogenes.

Cancer as a historically significant killer of juveniles. Are modern cancer statistics of use in getting at ancient cancer reality? Some facts suggest that they are not. We humans have dumped a lot of carcinogenic chemicals into the environment. And by conquering other diseases we've increased the number of candidates available for cancer death. But modern statistics nonetheless have value. After all, even if most cancers were caused by pollutants (which I doubt), humans and other animals are born with the initiating mechanism present in every cell. For that reason, facts about cancer's modern incidence, if used judiciously, can shed light on its role in the inaccessible past.

The most compelling statistic is cancer's overall kill rate. It now accounts for about twenty per cent of all human deaths in the United States and in other industrial countries. As already noted, it is the second leading killer of American children.

Cancer death rates increase dramatically with age, and the logic of selection tells us that those kinds of cancer that now kill the old once scourged the young. (Postponement of cancer's onset protected the germ line. To our genes, delay was as good as a cure.) Using that line of reasoning, no other figures are more convincing of cancer's past evolutionary potency than the startling statistics for prostate cancer in old men. According to Mostofi and Leestma, among men eighty years and older *80 to 90 per cent develop prostrate cancer.*

(I am not suggesting that our youthful nonancestors died of prostrate cancer in anything near those percentages. I don't need to. According to biologist A.F. Huxley, a new germ line mutation that gives a mere 1% advantage to its possessors will conquer a large population in about 100 generations. Although biology's number-crunchers apparently haven't come up with any comparable measure of how quickly selection eliminates lethal genes in juveniles, the logic would be the same. Any specific cause had to kill only a few juveniles per generation to change gene pools permanently and dramatically.)

Available cancer statistics for other species are compelling. Not long ago, two National Cancer Institute investigators [Anderwont and Dunn] examined 225 wild mice and found 121 tumors in 98-- an incidence of 43.6%. According to Warren Andrew, studies of fish in United States hatcheries disclosed tumor rates affecting in some cases "practically 100% of the population of the fish over 3 years of age."

Even if man-made carcinogens caused *all* those cancers, such extremely high incidences eliminate any doubt that primordial cancer had the *potential* to kill evolutionarily significant numbers (as low as Huxley's 1%) of juveniles.

And if cancer killed enough juveniles then many mechanisms that *delay* death, at least beyond the reproductive age, were accumulated. That brings us to that most startling (and illuminating) of all cancer phenomena, the "time bomb" phenomenon. According to John Cairns, a Harvard School of Public Health cancer specialist, *most human cancers take an extremely long time to start:*

> the interval between the first carcinogenic stimulus and the final appearance of the tumor tends to be *decades* rather than months or years...For example, after X-irradiation (either therapeutic or diagnostic, or resulting from exposure to the explosion from an atom bomb) the first cancer to appear is leukemia, which reaches its peak incidence about 7 years later...by contrast, the commoner kinds of cancer, like breast cancer, *do not start to increase in incidence for 10 to 20 years* but they probably remain high from then on. [Emphasis added.]

Cairns detects an even more dramatic delay in statistics for human penile cancer. Men who are circumcised in infancy have an incidence near zero. Men who are never circumcised enjoy no such exemption. That establishes that the disease is associated with the state of non-circumcision. But not all circumcisions are performed on infants; it is a coming-of-age rite in many societies. In those communities (Arab moslems and the Masai, for example) where boys are circumcised between the ages of 10 and 15, penile cancer occurs. But the cancer does not appear until old age--*fifty years* after the presumed latest possible initiation of the process! Cairns' correctly infers from these startling facts that the cancer starts in prepubescent boys and that the time bomb effect delays the onset of symptoms for five decades.

The significance of the time bomb effect as indirect evidence of heavy past genetic losses to cancer cannot be overestimated. Delay in the appearance of cancer symptoms is *not* in any way comparable to incubation periods for infectious diseases. The mechanism for cancer begins with damage to *self*, to a cell already equipped with a cancer trigger, not with invasion by an alien pathogenic micro-organism. Delay in the appearance of cancer symptoms has only one possible explanation: the defenses of the body interfere with the malignant process with *one-hundred percent effectiveness* for incredibly long periods--more than 18,000 days when symptoms do not appear for fifty years. And the appearance of symptoms decades after the incident informs us of another significant fact: cancer's imperative to kill never relented. Throughout the delay period, the initiation mechanism kept responding, possibly daily, to the decades-old mutational event. And, for the same incredibly long period, the defenses kept suppressing cancer. That sort of defense capability against an otherwise invincible killer has only one evolutionarily logical explanation: *uncountably large numbers of nonancestors in our lineage died of cancer.*

The time bomb phenomenon occurs in other species, with the length of the delay adjusted to allow for differences in life spans. Four-month-old hamsters inoculated with a carcinogen developed tumors two years later. When hamsters are two they are in their dotage, well beyond the age of reproduction.

Those patterns confirm past selection pressure in favor of antioncogenes. They confirm cancer-driven evolution.

The lethality of sunlight and the existence of defenses against it. I've pointed out that all invertebrate animals--which have no apparent active secondary defenses against cancer--avoid exposing unprotected somatic cells to sunlight. The vertebrates--which have cancer-specific immune systems--have dispensed with many first line defenses; many tolerate sunlight exposure. I attribute these phenomena to cancer selection.

Modern experience with melanoma is pertinent. This lethal skin cancer, caused by exposure to sunlight, is prevalent in Australia and in North America among people with fair skin, the kind that never tans, or tans only after burning. Most victims are of Celtic origin, the descendants of Irish and Scotch migrants. Moreover, melanoma incidence seems clearly related to the *amount* of sunlight exposure. One study found the incidence among Scots in sunny

Australia *one-thousand times higher* than among those who remained in cloudy Scotland.

Another relevant study compared the incidence of all types of skin cancer in four American cities. Investigators compared meteorological records of the relative sunlight in each city to the incidence of skin cancer per 100,000 inhabitants. The results, as reported by Groff Conklin in *Cancer Biology* showed that Detroit, with the lowest amount of sunshine (25 to 40 percent of the days were sunny) had the lowest rate of skin cancer, 24 cases per 100,000. Other cities showed a consistent correlation of days of sunlight to cancer cases: Pittsburgh, 50 to 57 percent and 37 cases; New Orleans, 62 to 64 percent, 129 cases; and Dallas with 60 to 80 percent sunshine, 140 cases.

Adding to the compelling case for sun-caused cancer are victims of *xeroderma pigmentosum,* a rare skin disorder. These people lack the DNA-repair mechanism found in normal cells, a device that fixes DNA damaged by exposure to *normal* levels of sunlight. Victims cannot tolerate *any* sunlight. If they go out-of-doors during daylight hours they are likely to get skin cancer at the sites struck by sunlight.

Albinos, who lack pigmentation, also experience high levels of skin cancer.

If we ask the "why did they bother?" question about either pigmentation or cellular DNA repair mechanisms we get the identical answer. Our lineage selected those devices, or improved upon them, to avoid extinction by sunlight-induced cancer. If we apply the "cause of death" rule, we come to the same conclusion: pigmentation and DNA repair mechanisms prevail because organ-isms not equipped with them died of cancer caused by solar radiation.

I've identified vertical daily migration of transparent marine animals as a cancer defense. My conclusion presumed that sea water shields fish from carcinogenic solar radiation. It does. Recently (December 26, 1989), Eric Schmidt reported in *The New York Times* that researchers (at Brookhaven National Laboratory) who were testing the effects of exposing fish to different components of ultra violet radiation placed the fish in 10 gallon tanks. "The tanks, however, were filled with [only] two inches of water. *If the water is too deep, the ultraviolet wavelengths are distorted, altering the radiation's effect."* [Emphasis added.] The experiments were "successful;" the fish developed lethal melanoma. Those results are consistent with my conclusion.

I've mentioned hair as a defense against sunlight-caused cancer. In 1983 researchers conducted experiments with nude (hairless) mice who were exposed to a known carcinogen, ethylnitrosorea. The mice developed skin tumors at rates far above those in a control group. In his article Schmidt reported that "fish were chosen because the usual test animals, mice and rats, are less useful in skin cancer studies [because] *fur blocks some radiation and also hinders tumor growth..."* [Emphasis added.] My view that hair was selected, perhaps primarily, to protect against cancer is supported by the evidence.

The power of first-line defenses in invertebrates. The theory states that invertebrate animals never developed secondary defenses having the same level of efficacy as the vertebrates' immune systems. As a result, they rely on powerful first-line defenses against UV radiation and other naturally-occurring carcinogens. Research confirms that conclusion.

I infer from my theory that vertebrates would be more susceptible than invertebrates to radiation damage--*other than cancer.* The reason? Vertebrates got rid of the primitive whole-body defenses that were effective, not only as cancer-avoiders, but against *all* radiation damage. By selecting genes for the cancer-specific backup system the vertebrates became more vulnerable to other forms of radiation damage.

Laboratory experiments confirm my reasoning.

Researchers use the concept of "LD50" to measure the lethality of radiation on experimental animals. LD50 refers to the lethal dose that kills 50 per cent of the animals within thirty days. LD50 for warm-blooded vertebrates is reached when they are exposed to radiation equal to less than 1000 rads (r). Using that figure as a benchmark, it is clear that invertebrates tolerate much higher levels of radiation before half the animals are killed. As reported by Clyde J. Dawe, some mollusks exhibit an LD50 at levels ranging from 8000r to 20000r. Adult *Drosophila* can tolerate 64,000r before they are even sterilized, and higher levels before LD50 is reached. Those data conform with the idea that invertebrates rely on primitive whole-body protection against radiation.

An even more compelling test was carried out with larval *Drosophila*. Those experiments showed that radiation defenses were extremely low at the beginning of the insects' life, although they were already climbing sharply a few hours later. At 3 hours of age LD50 was reached at only 200r. At 4 hours the larvae could handle 500r, and at 7 1/2 hours LD50 was 810r. Those low tolerances confirm my idea cancer selection lies behind the insects' protection of their young in heavily-shielded locations. Both the low tolerance of larvae--which were artificially removed from their normal sheltered habitat--and the extraordinarily high resistance to radiation of adult insects* support the idea that their lineages endured heavy losses to radiation-caused cancer.

It seems to me that a theory of evolution ought to explain not only the existence of radiation tolerance but why the insects' is so much greater than the mammals'. After all those lineages survived in the same macro-environment, the land surface of our planet. Why the huge variation?

The puzzle, ignored by neo-Darwinists, is solved by my theory. Heavy whole-body defenses protected insects against all radiation-caused damage, both cancer and gross damage to tissue. But the vertebrates' immune system, which can kill cancer cells, cannot repair tissue damage. (Because neither lineage was routinely exposed to high levels of radiation [there were no uranium miners or radiologists in their primordial past] natural selection can be eliminated as the explanation.) Only the lineage's different responses (single-phase defense versus two-phase defense) to intense cancer selection explains.

---

*Not surprisingly, attempts to use radiation to kill insect pests have been unsuccessful.

The evolutionary role of immune systems. Medical evidence supports the idea of the immune system's surveillance function. I've already mentioned the higher incidence of certain rare types of cancer in persons with AIDS. Other groups that have lowered immune responses--the aged, persons whose immune responses were deliberately suppressed to permit organ transplants, and persons born with deficient immune systems--all experienced higher rates of detectable cancers that are rare in the general population.

Those facts appear to have led one cancer professional to make a serious error in logic. John Cairns, in his 1978 book, *Cancer: Science and Society*, notes that although persons with lowered immunity seem to experience higher rates of *rare* cancers, they have the more common cancers at rates similar to persons with functioning immune systems. According to Cairns, those observations refute the idea that the immune system serves as a watchdog against all cancer: "The result leads to the conclusion that most of the common human cancers are not subject to immune surveillance..."

Cairns errs. The facts he cites actually support the idea of universal immune surveillance against cancer.

His mistake stems from the classification of some cancers as "common" and others as "rare."

No one, including cancer specialists, can *see* what is going on at the cellular level in living animals. Because he cannot observe cancer initiation at the cellular level Cairns has no basis for characterizing different types of a disease that is subjected to *any* surveillance as either "common" or "rare". (Cairns acknowledges *some* surveillance when he notes that "rare" cancers increase when the system is impaired.) Using only indirect evidence, he has assumed that cancers that cause identifiable *symptoms* in persons with functioning immune systems are approximately representative of all cancers that are *initiated*, including those that are routinely extinguished by the immune system. His assumption is wrong.

Using a more cautious analytical approach, I come to precisely the opposite conclusion. I think it more prudent to presume that if the immune system works against *any* cancers *it would work best against cancers that were historically responsible for the greatest numbers of juvenile deaths.* After all, those would be the types of cancer on which the immune system cut its teeth, the ones that had most frequently killed juveniles. The immune systems would have had less experience against cancers that cumulatively have killed fewer genes.

In a population of persons with *normal* immune systems--whose evolution was influenced by cancer--which of those two groups of cancer would cause the greater number of deaths, the common cancers or the rare cancers? Obviously, the immune system would be less likely to stop those that have been historically rare; they would kill more victims. And those that were historically more common would cause the fewer deaths because the immune system would have long ago learned how to control them.

I am saying that cancers physicians perceive as "rare" are actually common (occurring often, but routinely snuffed out by the well-drilled immune system) and those that doctors perceive as "common" are actually rare (occurring less often but outwitting the immune system more often) in terms of the number of times in a typical lifetime that a particular type of cancer is *initiated.*

When the immune system is knocked out, as in AIDS victims, the true incidence of the various types of cancer *initiated* suddenly becomes clear. Those that are routinely terminated by healthy immune systems, like Kaposi's sarcoma, now appear with greatly in-creased frequency, while those rarer cancers (rarer, that is, in terms of initiation-incidence) that are *not* routinely eliminated by the immune system, continue to occur at or near the same levels as in the healthy population.

Then there is the problem of AIDS victims dying of something other than cancer. In 1983 one AIDS expert, Robert A. Good, af-ter noting that most AIDS victims succumb not to cancer, but to opportunistic parasite infections, related that fact to cancer in AIDS victims:

Whether other cancers that reflect deficiencies in immunological surveillance will appear will only be revealed in time if and when the opportunistic infections can be sufficiently well controlled to permit more prolonged survival and thus [provide] a latent period long enough to spawn other forms of cancer.*

That Good was on the right track was borne out in 1990 when the head of the National Cancer Institute reported, "Cancer is emerging as the biggest new challenge in the treatment of AIDS." This upsurge, reported in *The New York Times*, was attributed to the introduction of drugs that enabled victims to live longer and to avoid death from pneumonia, the original major killer of AIDS victims.

Both the growing increase of cancer and the abnormal mix of types of cancers in AIDS victims support the idea that our immune system routinely surveills against cancer.

Moreover, other data supports the same idea.

Judah Folkman, in *Cancer Biology*, notes that about 1,600 new cases of cancer are diagnosed every day in the United States. Although that seems like a high number, Folkman puts it in proper perspective. Each case of cancer begins in one cell that normally divides inside our bodies. The total number of cells dividing every day in the United States is about 10^{20} (1 with 20 zeroes following it). Using those two facts, Folkman estimates the probability of cancer occurring as "extremely small, on the order of 10^{-17}."

In the same volume, the editor, Errol C. Freidberg, adds a perceptive comment about Folkman's statistics.

We know from studies in tissue culture that the frequency of neoplastic transformation of individual human cells is significantly greater than 10^{-17}. *Hence it is likely that many, perhaps the vast majority, of neoplastically transformed cells in vivo do not progress to become clinically diagnosable.* [Emphasis added.]

If the transformation of individual cells to the cancerous state is significantly higher than diagnosable cancer, then *something* is preventing the transformed cells from killing more victims or even progressing to the "clinically diagnosable" stage.

Finally, researchers are currently working assiduously to develop cancer treatments based on enhancement of the immune system's ability to kill cancer cells.

In short, all available data, if correctly interpreted, support my assertion that immune systems enabled vertebrates to risk higher levels of cancer initiation.

The cancer-defensive nature of sleep. Although the important facts already mentioned--cell division during sleep, sleep's occurrence only in animals with immune systems, increased sleep when cell production is at its highest, growth hormone secretion during sleep and the avoidance of sunlight during sleep by most animals--support the idea that sleep originated as a defense against cancer, I am not aware of any direct evidence that would shed light on that proposal. Future discoveries of increased immune activity or cell repair during sleep would support my theory.

Nondividing cells functioning as a shield against cancer. My idea that selection favored the placement of muscle cells and nerve cells (including those of the human brain) between the normal source of UV radiation (the sun) and the cancer-vulnerable pre-mitotic cells is--as far as I know--original. I don't think it has been tested, but I have thought of an experiment that would do just that.

Hairless mice (they must be nude to eliminate the cancer-defensive nature of the fur) would be placed in cages and UV radiation would be directed at them *from beneath the cages.* Placing the source of UV radiation under the animals would circumvent whatever physical protection muscle and nerve cells afford the animals' dividing cells. It would be equivalent to turning the mice upside down. A control group could be subjected to the same dosage of UV radiation, but, in emulation of normal conditions, from an overhead source. A higher incidence of tumor formation in the first group over that experienced in the second group would be consistent with my assertion that nondividing cells provide protection from sunlight-induced cancer.

Regeneration Avoidance as a Cancer Defense. Certain human cancers are clearly associated with the increased cell renewal activity required to repair damaged tissue. They support the idea that cancer selection caused early animal lineages to reject genes for aggressive regeneration.

In India, men whose inner thighs are constantly chafed by loin cloths can develop cancer at that site.

Ill-fitting dentures, which continually injure gums thereby initiating excessive regeneration, can lead to cancer. So can peptic ulcers, open sores in the stomach's lining.

Asbestos, a notorious carcinogen, is a chemically inert substance and inert substances, by definition, are not mutagenic. Under a microscope, however, asbestos can be seen to consist of many jagged-edged fibers. Any fiber that punctured lung cells would, because of those rasp-like protrusions, remain in the tissue and tear away at cells. This repeated trauma would cause continual cell regeneration and heightened risk of cancer.

More evidence that regeneration can lead to cancer was reported (in *Science*, in 1983), by Michelle Goyette and her associates at Brown University. After observing damaged liver tissue in rats, these scientists concluded: "The number of [copies] of...cellular oncogenes...increases during liver regeneration..."

Those findings are consistent with my idea that cancer selection worked against regeneration of the sort common in nonanimals.

The absence of cancer in nonanimals. Nothing is more offensive to evolutionary logic than the suggestion that plants can get cancer. Organisms that have depended on continual direct exposure to a powerful carcinogen for many hundreds of millions of years are telling us, emphatically, that they can't get cancer.

Nature conforms with logic. Cancer has not been found in plants.

Some older cancer texts carelessly refer to crown gall disease, a disorder found in trees, as "plant cancer", but it is nothing of the sort. Crown gall disease has never been known to kill a tree and its cause has been identified as a bacterium. Contrast those facts with cancer: untreated, it is nearly always fatal, and any possible causative role for bacteria was looked at and discarded long ago as unworthy of further research.

Experimenters have exposed plants to other forms of radiation--X-rays and gamma radiation. Those carcinogen-mutagens caused somatic mutations (in one experiment they turned to white the petals of a red Dahlia), but not cancer. That is precisely the kind of reaction we ought to expect in multicells that are unequipped with functioning oncogenes in every cell: noticeable somatic changes caused by mutated genes, but no cancer.

In 1986 Esther C. Peters and her colleagues reported neoplasms, cancer-like growths, in stony coral growing off the coast of Florida.* The researchers were unable to identify the causative agent. If those growths were caused by a virus that inserted foreign (i.e., non-coral) DNA into the nucleus of a cell, then the neoplasm did not occur in a "pure" coral cell. The DNA imported by the virus could have originated in an animal cell and could have contained its own oncogene. If that happened, the growths would not be a basis for suspecting that cancer caused by environmental factors, especially solar radiation, played a significant role in the evolution of coral.

In considering the indirect evidence, however, it must be acknowledged that unlike most cnidarians, which regularly expose themselves to direct sunlight, modern coral's relationship with sunlight seems ambiguous. All coral reefs grow heliotropically, and coral that were artificially deprived of light in an experiment died within 18 days. Those observations suggest dependency on direct exposure to UV radiation, which would be inconsistent with a history of cancer selection. (Zooanthelae, algae that require

---

*In a private conversation with the author, a scientist (whose name does not appear in this book) expressed doubt about the diagnosis of neoplasm.

sunlight for photosynthesis, live inside coral reefs and, among other things, get rid of carbon dioxide produced by the coral.) Howeverer, the coral organisms themselves seem to avoid sunlight; they emerge from behind their stony exoskeleton only after dark.

Although it seems unlikely that the lineages that produced these simple, long-lived creatures were subjected to any cancer selection, if future investigations clearly demonstrate that cancer can be repeatedly *initiated* in stony coral by solar radiation (applied at intensity levels found in nature) then my theory would require modification, but it would not be refuted. Deliberate cancer-initiation in *other* nonanimals would legitimately refute the theory and I make a specific proposal along those lines in Chapter Thirteen.

Although the modern cancer facts I summarize in this chapter support the theory, no theory of evolution can be evaluated solely by consideration of physical evidence. Many parts of my theory, including the cancer-defensive origin of strict genetic control over the production of somatic cells and its role in the origin of phenotypic uniformity, bilateral symmetry and aging can be evaluated only if the evaluator correctly assesses the logic of the theory, including its parsimony, resonance, and explanatory power.

Twelve

Is It Science?

The great power of this principle of selection is not hypothetical.

Charles Darwin

Authors should distinguish between, on the one hand, new
analyses and concepts that are based upon established data or
the existing body of biological theory and, on the other hand,
purely speculative suggestions that are not so based. The
latter will not be acceptable for the Journal.

Journal of Theoretical Biology

    Can a book written by a man with virtually no formal scien-
tific training and who has never been employed by a university, a
museum or a laboratory, be scientific? A book written not in
scientific jargon, but in language understood by most general readers? One that contains no mathematics and only a few diagrams?
A book whose central idea cannot be directly tested in a laboratory?

    That's what some people asked in 1859 when Charles Darwin
published *The Origin of Species*. Darwin had *not* been educated
as a scientist--"I consider all that I have learned of any value has

been self-taught."* Only 27 years old when he completed his tour as a naturalist on *HMS Beagle*, he never again worked as a professional scientist. The *Origin* contained no mathematics and just a few diagrams. He wrote it for the general public. (It was an instant best seller.) His central claim could never be tested; the origin of no species has ever been observed and probably never will. For all those reasons critics said Darwin's work was not scientific.

Some may raise similar objections to my work. I have had no scientific or medical training. Worse, unlike Darwin, I've never been employed in *any* scientific capacity. And yet I claim, of all things, superior conceptual knowledge of cancer, perhaps the most intensely studied and most intractable of all medico-scientific problems, and of evolution, the central theory of biology. I use no mathematics and very few diagrams. I've done no original research. (Darwin was a prodigious researcher.) I don't examine fossils, I don't study living animals and I've never so much as looked at a cancer cell through a microscope. Why should anyone take me seriously?

Well, the skeptics were wrong about Darwin--the *Origin* is science of the highest caliber--and they will be wrong about me. ‸

The question of whether or not the originator of an idea has formal scientific training is, of course, a non-issue. Ideas should be judged on their merits, not by their author's credentials, or lack of them. That is especially true of evolutionary theory for there isn't a university in the world that even *pretends* to issue credentials in evolutionary theory development or, for that matter, evolutionary theory evaluation. All the so-called professional evolutionists were trained and accredited in other fields, such as population genetics, zoology, botany and paleontology. And only a few of the many thousands of scientists credentialed in those and other biological specialties try their hand at evolutionary theory. In other words,

---

*Alfred Russel Wallace could have uttered the same phrase. According to Edey and Johanson, "Everything [Wallace] learned--optics, mathematics, surveying--he learned by himself..."

professional theorists are, like me, self-taught and self-appointed.
There is a more important reason not to use my lack of credentials as a reason for dismissing my idea. As all well-informed students of the history of science know, it is not at all unusual for outsiders to come up with solutions that elude professionals. In fact, it has occurred so often that it could be considered the norm, not the spectacular exception. Thomas S. Kuhn in his classic work, *The Nature of Scientific Revolutions*, put it succinctly:

> *Almost always* [persons] who achieve these fundamental inventions of a new paradigm have been either very young or *very new to the field* whose paradigm they change. [Emphasis added.]

According to Kuhn, professional scientists are not only *not* trained to develop new theoretical concepts, they have all been indoctrinated to avoid the one style of thinking--deductive reasoning--that any revolutionary thinker must employ:

> [Scientific training] is a narrow and rigid education, probably more so than any other except perhaps in orthodox theology... [It] is not well designed to produce the [person] who will easily discover a fresh approach.

Kuhn even suggests that people who devote their lives to particular fields of study are the *least* likely to make important discoveries in their own field.

> Sometimes a normal problem, one that ought to be solvable by known rules and procedures, resists the reiterated onslaught of the ablest members of the group within whose competence it falls.

Perhaps that is why the armies of scientists who investigated cancer failed to see that it had all the characteristics of a powerful evolutionary function.

German author C.W. Ceram, in *Gods, Graves and Scholars: The Story of Archeology*, wrote at length of the remarkable role outsiders have played in science:

No matter how far back we go in the history of science, it seems that *an extraordinary number of great discoveries were made by dilettantes, amateurs, outsiders*--the self-taught who were driven by an obsessive idea, unequipped with the brakes of professional training and the blinkers worn by the specialists, so that they were able to leap over the hurdles set up by academic tradition. [Emphasis added.]*

Ceram cites many examples of outsiders succeeding where the professionals failed:

Otto von Guericke, the greatest German physicist of the seventeenth century, was a jurist by profession. Denis Papin, eighteenth-century pioneer in the development of the steam engine, was a medical man. Benjamin Franklin, son of a soapmaker, without even a secondary education, not to mention university training, became not only a great statesman but a scientist of note. Luigi Galvani, the discoverer of electricity, was another medical man who owed his discovery...precisely to the deficiencies of his knowledge in the field in which he made it. Joseph von Fraunhofer, the author of a distinguished work on the spectrum, could not read or write before he was fourteen years of age. Michael Faraday was the son of a smith, a bookbinder's apprentice, and almost completely self-educated. Julius Robert Mayer, who discovered the law of conservation of energy, was a physician, not a physicist. Another physician, Hermann L. F. von Helmholtz, published his

---

*From GODS, GRAVES AND SCHOLARS by C.W. Ceram, translated by E B Garside. Copyright 1951 by Alfred A. Knopf, Inc. Reprinted by permission of the publisher.

first work on the same subject at the age of twenty-six. Georges, Comte de Buffon, a mathematician and physicist, published his most significant work in the field of geology. The man who built the first electric telegraph was a professor of anatomy, Thomas Soemmering. Samuel Morse was a painter, as was Louis Daguerre; yet the first created the alphabet for the telegraph, the second invented photography. The fanatics who invented the dirigible, Ferdinand Count Zeppelin, Gross, and Parseval, were military officers without a trace of technical training. The list is endless...

Two nineteenth century science giants Ceram might have added to his long list are Charles Lyell, the father of modern geology, whose theorizing influenced Darwin, and Gregor Mendel, whose brilliant experiments laid the foundation for the science of genetics. Lyell had been trained in law and Mendel was a priest. And in the twentieth century we have had the marvelously instructive example of Alfred Wegener and his theory of continental drift.

I once made the mistake, while at a cocktail party in the late 1950s, of asking a friend, an oil-field geologist, what he thought of Wegener's theory of continental drift. I had recently read, in a popular magazine, about Wegener's controversial and radical proposal. He had suggested--in 1912--that all the continents originally comprised a single land mass that had broken apart and drifted to their present locations. The then-conventional view was that, although land bridges may have connected the continents in the past, the continents themselves had always been located exactly where they are now.

I had found Wegener's reasoning compelling: he showed that the coastlines of the present continents all fit together like pieces of a planet-sized jigsaw puzzle. The only logical explanation for that fitness, he argued, was that they had broken off from a single mass and moved to their present positions.

To my surprise, my geologist friend, normally an amiable fellow, became rather annoyed by my question. "That ridiculous idea has been around for years! It's utter nonsense! Wegener was nothing but a *meteorologist* who didn't understand the first thing about geology. He should have stuck with isobars and weather forecasts!" Those were not my friend's exact words, for I no longer remember them, but he was scornful and dismissive of Wegener's radical idea.

My friend was wrong. All geology professors now teach their students that the continents were indeed originally part of a single land mass. Wegener, the outsider, the "mere" meteorologist, had been several decades ahead of the contemptuous professionals.

The geologists' error was in conflating Wegener's incorrect guess that tidal force had moved the continents with his correct inference that they were once one. The matter was not settled until the 1960s. That's when movable tectonic plates--which *can* move continents--were discovered and Wegener's basic idea was universally accepted.

Ceram writes extensively about yet another rank outsider, Heinrich Schliemann, a nineteenth century businessman and amateur archeologist who searched for ancient ruins where the professionals said he would find nothing. Schliemann, who had the wit to use the Iliad as a guide, did what the professionals had only dreamed of doing; he found the ruins of ancient Troy. Later, again using Homeric evidence, he discovered the site of the ancient Cretian capital of Mycenae. But despite those great accomplishments, perhaps the most spectacular in the history of archeology, Schliemann earned nothing but contempt from the academics during his lifetime. According to Ceram, that reaction is common, and has its roots in a rather unappealing tendency: "The professional's mistrust of the successful outsider is the mediocrity's mistrust of the genius."

When I began the work that led to publication of my theory in the *Journal of Theoretical Biology*, I proceeded, without giving it any thought whatsoever, on the presumption that mathematics could contribute little to the development of theories of evolution. Fortunately, I had not yet read *Evolution,* a scientific journal that purports to publish articles about evolution. In issue after issue its pages are crammed with equations, charts and diagrams. It is simply saturated with data and mathematical symbols. In spite of the popularity of algebra, calculus and other forms of mental iron-pumping in it and other biology journals, however, a few thoughtful evolutionists had already come to the same sensible conclusion I had reached on my own; they don't think mathematics are of much use either. Michael T. Ghiselin of the California Academy of Sciences addressed the question in *The Triumph of the Darwinian Method:*

> there are those...who maintain that scientific work must employ mathematics. But mathematics may reasonably be treated as a branch of logic, and to view any form of logic as something more mysterious or valid than what is called common sense is without foundation and smacks of the superstition of numerology. Scientific inferences should be accepted because the premises are true and because the conclusions follow logically. The truth does not derive from the jargon in which it is expressed.

Ernst Mayr, in *The Growth of Biological Thought*, commented at some length on the "blind faith in the magic of numbers and quantities" held by many non-biologists and "the myth that mathematics [is the] queen of the sciences." He approvingly cites Pierre Bayle, a seventeenth century scientist, who argued that "historical certainty was not inferior but simply different from mathematical certainty" and Buffon, who observed that some subjects--most emphatically including evolution--are simply "far too complicated for a useful employment of mathematics".

Mayr is equally unenthusiastic about the usefulness of experiments:

> No one questions that the appropriate technique for the study of functional phenomena is the experiment; but it must be emphasized that the explanation of historical [evolutionary]

phenomena must rely on inferences from observations. The blindness of many experimentalists...[is] caused to a large extent by their blind insistence that only the experiment can give answers to scientific questions...When part of a historical narrative consists of functional processes, they can be tested by experiment. But the historical sequence as such...can be reconstructed *only on the basis of inferences derived from observations.* [Emphasis added.]

I will put my views on the severely limited value of experimentation (and its close relative, field observation of living organisms) more forcefully, even rudely, for there is a grave, hidden risk in all biological observations, one that persons who work in the inorganic sciences do not face. Simply put, it is both arrogant and stupid to assume that a *cumulative* transformational process would provide researchers with *any* representative samples of the selective events that took place over the very long period in which that process was operating. Indeed, the very nature of selection *guarantees* that no examined sample of biological events can *possibly* typify the universe of events that occurred in the lineage. Selection--natural selection as well as cancer selection--eliminated or minimized many threats to the germ lines' existence long before the animals studied by the modern observer came to exist.

Because past selection worked on "then" problems the most evolutionarily significant events that took place in the past cannot possibly recur in "now" populations.

For that compelling reason, attempts to extrapolate against the arrow of time, from observed biological events to those in the past, run the risk of simplism and error. Anyone who assumes, to cite a highly relevant example, that the low incidence of cancer observed in present-day insects is a reason for doubting that cancer selection was a major factor in their evolution is in the wrong science. In fact, precisely that error has already been made and immortalized by the very founders of population genetics.

Those scientists (Fisher, Haldane, Morgan, Muller, Dobzhansky and others) whose work in the 1920s and 1930s is revered by present-day biologists as essential to the formulation of neo-Darwinism, all assumed, consciously or unconsciously, that because they did not see cancer in their fly rooms and didn't need it to explain the minor transformations they observed, it had nothing to do with the evolution of *Drosophila melanogaster,* and, by extension, with the evolution of any animal.

That "If-I-can't-see-it-in-my-laboratory-it-didn't-happen" attitude may be valid in physical sciences or when applied to here-and-now biological problems, but it is utterly inappropriate to the problem of historical selection.   Macho-skepticism may enable its practitioners to feel superior to others, but they are really exhibiting gross scientific ignorance. Induction-based conclusions--generalizations developed from observations of phenomena--are appropriate and powerful tools in establishing a foundation for predicting *future* events, especially those in the near future. But unless used with great care, the application of inductive methods to the biological past is fraught with risk. It can easily lead--as it did in the "we've identified-all-evolutionarily-important-biological-mechanisms"conclusion--to embarrassing intellectual pratfalls.

Despite the great risk of error inherent in observational science, many professionals act as if it is the *only* way to solve evolutionary problems. *Evolution* seems to publish nothing but "show and tell" reports--most of them dressed up with arrays of diagrams and models or battalions of mathematical symbols--in which researchers share with the reader their observations of present day organisms. They seem not to understand that evolution happened in the irreproducible past.

If observation, mathematics and experimentation are of little use in evolutionary biology, can we scientifically determine whether or not cancer selection played a major role?   The answer is a resounding yes. There *are* valid scientific methods that enable us to "get at" events in the deep past. Some are used routinely by cosmologists, the scientists who specialize in the origin of the universe. They can reach *scientific* conclusions about events that occurred 15 or 20 billion years ago. And because they deal with elemental matters that are much less complex than biological entities, cosmologists can use mathematics extensively. But when it comes to probing the mysteries of early biological evolution the problems are substantively different. Listen to Mayr:

> biological evolution is fundamentally different from cosmic evolution. For one thing, it is more complicated than cosmic evolution, and the living systems that are its products are far more complex than any non-living system...

Even the simplest organisms--viruses and bacteria--are far more complex than any inorganic molecules. Moreover, biology differs from physical sciences in that much about its past can never be determined with exactness. It would be foolish, for example, to expect paleontologists to find fossilized remains of the founding species of the animal kingdom. Transformational evolution has wrought such great changes that those animals can only be described in a general way. Nor can we expect to find in the fossils a direct record of cancer deaths.*

But just because we can never measure past cancer deaths does not mean that attempts to determine if such deaths were evolutionarily significant are unscientific. Or speculative. Inability to measure is a fundamental feature of evolutionary biology. Evolutionists must work with a broad brush, not a sharpened pencil. Cosmologists may confidently talk about how many seconds after the Big Bang hydrogen and helium came to exist, but biologists can

---

*The only thing known about the reliability of the fossil record is its unknowability; we have no way of figuring out how representative it is. For those who relish mathematical explanations, if you know the numerator (the fossils) but not the denominator (all the animals that ever lived) then you cannot even estimate the value of the fraction. Having said that, for what it's worth, bone tumors have been found in dinosaur fossils and Niles Eldridge, a paleontologist, claimed (1982) to have discovered a fossilized tribolite "with [a] growth in its left axial furrow" that looked to him like a tumor.

only estimate, with a margin of error that may exceed 100 million years, how soon after one biological Big Bang (the origin of multicellular life) a second Big Bang (the beginning of cancer selection) occurred. But the inability to measure past events does not alter the reality of those events. What happened, happened.

It is only by correct application of the methods I have used--conceptual, nonmathematical and deductive--that general scientific knowledge of those past events can be grasped. And if the scientific methodology employed is sound, then the knowledge gained is valid.

Now I will show that the method I used is consistent with one of our greatest scientist's ideas on the construction of scientific theories.

Albert Einstein wrote one of the most lucid and succinct explanations of how a scientific theory is actually constructed--a description of the reasoning actually used by successful theorists. It was summarized and explicated by Gerald Holton in the Summer 1979 edition of *The American Scholar* in an article entitled "Einstein's Model."

Although Einstein was referring to his own work when he developed the model, I was reassured to learn from Holton's article that I had already used the method described by the great physicist in developing my new theory of evolution. (The main elements of my theory were in place by mid-1978.) I was pleased, but I was not surprised. Theory-building *ought* to be the same regardless of the science.

I have adapted (see Figure 3) the schematic used by Einstein, and amplified by Holton, to illustrate the process I used--and to demonstrate its scientific validity.

The model is best understood by starting where any theorist starts--with the evidence he can actually perceive with his own senses. Using the knowledge he possesses the theorist (T) makes an original guess--no other word is appropriate--about the connectedness of the phenomena he is observing. That guess is shown by the arc marked A. (My initial guess was that cancer was an anti-mutation mechanism.)   Starting with the guess, the theorist constructs a situation (shown as S) that he thinks is correct. He then draws, by reasoning deductively, certain inferences of what, logically, would have happened *if* the situation he hypothesized were correct. In the model I have identified those inferences as I^1, I^2, I^3, and so forth.

FIGURE THREE
EINSTEIN'S MODEL

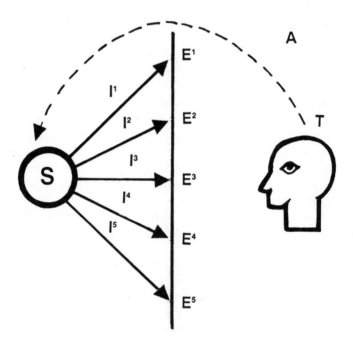

Einstein described that process--where "inference follows inference"--as requiring "much intense hard thinking" to ensure that all the appropriate conclusions are identified and thought through correctly and that they all flow logically from the postulated situation (S). (Einstein's comment added to my confidence: developing my own inferences was the most demanding intellectual work I have ever undertaken.)

The development of inferences is the heart of the hypothetico-deductive method--or the "if-then" process--that Einstein incorporates in his model as the S-I sequence. *It is the only method that can be used to develop scientific theories for phenomena that cannot be observed directly.* It was used by Copernicus, Newton and Einstein. It was used by Darwin and Wallace to develop their theory of evolution by natural selection. And I used it to construct the successor to their theory.

In the next step of theory development the theorist compares his individual inferences (I^1, I^2, etcetera) to observable phenomena--to real world evidence. Each piece of evidence conforming with individual inferences is identified in the model as E^1, E^2, and so forth. For example, I inferred, based on (1) my presumption that there were oncogenes in all animal cells and (2) the known carcinogenicity of sunlight, that the early animals would have avoided exposure to sunlight. The fossil evidence (E) supports, overwhelmingly, that inference.

Einstein pointed out that the more inferences (I) the theorist can *correctly* draw from the situation (S) that are supported by evidence (E), the more confident he becomes that the situation--and thus the entire theory--is correct. Holton explains, "...a theory gains more and more plausibility or usefulness the longer the predictions derivable from it are found to correspond to the growing area of available sense experience--and the fewer the contradictions."

The theorist revises the situation (S) when the evidence (E) demands it. In my own case, I had to face the fact that many modern animals expose unpigmented cells with impunity. That led me to investigate the phylogenic history of the immune system and to expand the situation (S) to reflect the hypothesis that vertebrates had cancer-specific immune systems and invertebrates did not. After much intense thought, I made additional inferences about the awesome influence of the immune system.

The process includes a feedback mechanism. This enables the theorist to correct mistakes. If the theorist discovers new evidence he may decide to alter the content of S, the situation, to allow for the new facts. To cite an example from my own experience, I made a very serious error in logic at the earliest stages in developing my theory; I erred in drawing inferences (I) from the situation (S). Specifically, I was attempting to figure out, by analyzing their characteristics, which organisms had been created by lineages that had cancer triggers in every cell. For various reasons, all of them terribly wrong, I concluded that the insect lineages could not possibly have experienced cancer. Shortly afterwards, a scientist to whom I had sent a draft of my work brusquely informed me that cancer had already been found in insects. That news--in Einstein's schematic, evidence (E)--completely wrecked my theory as it then existed.

I went back to work and revised the situation (S) to include the insects among those organisms whose existence was largely determined by cancer selection. I had to rethink my inferences, especially those about the nature of cancer defenses. Those revisions fundamentally changed--and improved--the theory.

The feedback mechanism worked for me because I was convinced that my basic idea--that cancer selection was a powerful evolutionary function--was correct. In its final published form the theory was plausible, coherent and consistent with known evidence. I may be a self-taught amateur, but I used correct *scientific* methodology.

Although I was convinced of the correctness of my theory long before I came across Einstein's model, it was comforting to learn that the great scientist and I operated on the same premise: pure reasoning *can* get at the truth.

I found Einstein's model useful in developing strategies for *at-*

tacking scientific theories as well as in constructing them. It identifies three points at which a critic can attempt to destroy mine or any theory of evolution.

The first is to consider whether or not the situation proposed is plausible. For example, the proposal that the moon is made of cheese is completely lacking in plausibility and no sane person would bother to try to falsify it.

The second point of attack is the inferences drawn by the theorist. The S-I step must stand up to rigorous analysis regardless of any correlations of I with E, for the simple reason that it might be possible to have an idea that conforms with reality in spite of it being based on erroneous logic. If I were to propose that heavy consumption of peanut butter causes children to speak English with American accents, I could point to two unimpeachable pieces of evidence in support of my proposition: (1) children living in the United States eat a lot of peanut butter and (2) they speak English with American accents. But there's no *logical* reason to think that consumption of certain foods determines accents. The "evidence" is nothing of the sort.

Contrary to that imagined example, the inference I drew from the carcinogen-mutagen correlation--that selected defenses against cancer would have enhanced precise development--is logically correct. Any successful attack on my theory must defeat that logic.

Rigorous evaluation of logic is especially important in evolutionary biology because it is only the (logic-dependant) inferences derived from the hypothesized situation--*not the hypothesis itself*-- that can be tested.

That idea conflicts with what high school teachers--incorrectly--taught us is *the* scientific method: develop a hypothesis, they instructed us, then test *it* in an experiment. That misconception-- that "chemistry set" science is the *only* valid method--leads some people, including many professional scientists (who ought to know better) to conclude that evolutionary theories are not scientific. Of course they're scientific. It's just that both their development *and their evaluation* demand deep thought and extended chains of reasoning.

Unfortunately, many professional scientists erroneously dismiss as speculative any idea that requires prolonged reasoning to evaluate.* But evolutionary theory is different. It *demands* thoughtful evaluation. That may not please the "chemistry set set," but that's the way it is.

The third means of attacking a theory of evolution is to apply Karl Popper's falsifiability test, to uncover evidence that refutes the inferences. Refutation of the inferences would, if they were arrived at logically, also refute the situation. The entire theory would crumble.

Any idea that can never be disproved--the idea that God created the Universe, for example--is not scientific. The ideologies of Freud and Marx, despite their lingering popularity in some quarters, are also not scientific because they can never be subjected to the Popperian test of refutation. But both my theory and neo-Darwinism are falsifiable.

I will use Einstein's model to recapitulate my refutation of the old theory. However, before doing so I will comment on two tests of the old theory that have been advanced by others and explain why they do *not* really test it.

The first erroneous test was proposed by Charles Darwin himself. He suggested, in *The Origin of Species*, that his theory of evolution by natural selection would be disproved if someone found an organ in one species of animals that existed solely for the benefit of another species. Now that proposed test would be adequate if Darwin was merely claiming that his was a scientific theory and that the old idea--that God created all life forms--was wrong. But the discovery of any such organ would *not* refute his central claim--which I reject completely--that natural selection was a *sufficient* explanation for animal evolution. It would refute quite a different

---

*Even some professional evolutionary biologists commit this error. A number of them told me that my work was speculation prior to its publication in a journal that has a firm policy of not publishing speculation. Implicitly, those scientists held that their belief (and it is nothing more) that cancer selection had nothing to do with evolution, was unchallengeable scientific truth even though it had never been seriously examined.

idea, one whose validity I fully accept. It would refute the proposal that natural selection played a *necessary* role in animal evolution. Professor John Maynard Smith of the University of Sussex, a resolute supporter of neo-Darwinism, has devised another so-called test of the scientific validity of the old theory. Like Darwin, he suggested (in his 1972 book *On Evolution*) an imaginary test. He said "if someone discovers a deep-sea fish with varying numbers of luminous dots on its tail, the number at any one time having the property of always being a prime number" he would regard it as "rather strong evidence against neo-Darwinism." His idea is that there could be no adaptive reason why the spots always equalled primary numbers. He went on to say that if those fantasized spots displayed replicas of the astral constellations it would totally demolish the theory.

Maynard Smith commits the same error as Darwin. The only conclusion one could reach if that imaginary fish were encountered is that some supernatural intelligence must have placed those strange dots on its tail. Like Darwin's imaginary parts that benefit another species, Maynard Smith's argument is appropriate in dealing with claims that evolution (which is *not* the same as evolution caused by natural selection) did not happen, that a Creator supervised the origin of life. But neo-Darwinism does not stop with that modest claim. It proclaims that natural selection is, of itself, a *sufficient* explanation for fish (and all the other animals) even if they have no spots on their tails. Like Darwin's, Maynard Smith's proposal does *not* test that proposition.

But the old theory can be falsified *in principle* (because it is scientific) by falsifying it *in fact* (because it is wrong).

Using the model in Figure 3 as a guide we can state the old theory's situation (S) as follows: once sexually-reproducing multicellular organisms came into existence they evolved to their present state largely as a result of natural selection--and (implicitly, but emphatically) *cancer selection had nothing to do with it*. We can then move to the inferences that might be drawn from that proposition and see if they meet the tests of plausibility and consistency with reality.

One of the first inferences from the old theory's situation--I consider it the primary inference--is a prohibition: There cannot exist any nontrivial feature in any organisms formed (supposedly) by natural selection that cannot be explained as having been selected because it added to the ability of the DNA inside that organism to survive in an environment in which natural selection (but not cancer selection) functioned. In other words, if the old theory is correct then nothing of significance can exist that cannot be explained by natural selection alone.

If the organisms we choose are not animals, then the old theory stands up to the test. I know of no major character in plants, jellyfish or sponges that is not sufficiently explained by the theory of evolution by natural selection.

But when we consider the animals, the old theory encounters enormous problems. *The inferences (I) we can plausibly draw from the situation (S) that neo-Darwinism postulates are not consistent with the evidence (E) provided by living or fossilized animals.* As I've been saying, all animals are replete with features--major characteristics, not trivialities such as pinhole-sized apertures found in the shells of some snail*--*which could not exist if they had evolved solely by means of natural selection.* The uniformity of conspecifics' phenotypes (to the point of eutely in nematodes), strict genetic control over the cell-by-cell construction of animal bodies, bilateral symmetry, genetically-controlled aging, the long history of universal sunlight avoidance, and deadly oncogenes in every cell--not one of those *major* phenomena, found in *all* animals, is consistent with the principal inference logically drawn from the situation postulated by the theory of evolution by natural selection.

Without my radically different presumption--that cancer selection was a major factor in animal evolution--none of those fundamental characteristics would have given the earliest animals any survival advantage whatsoever. In fact, some of them--cellular

---

*I once sat through a lecture by a well-known biologist who proclaimed that the existence of such holes had evolutionary significance. I comment further on this in Chapter Sixteen.

oncogenes and genetically-initiated aging--are downright lethal; rigorous application of the old theory's mechanisms tells us that they ought not to exist.

If the theory of evolution by natural selection cannot explain the fundamental properties of animals--and it cannot--then the theory is refuted. It is refuted, not by imaginary, nonexistent organs or by fanciful arrangements of luminous spots on fantasized fish--but by the fossils of tribolites, by the fly on the wall and by the reader of this book. *It is refuted by every single animal that ever existed.* Because the principal inference drawn from the theory is overwhelmingly contradicted by the evidence, the postulated situation on which the inferences are based is wrong. The theory *is* error.

Neo-Darwinism deserves to be treated as a serious scientific proposition (or theory) because it is falsif*iable.* Unfortunately for its advocates, it is falsif*ied* by the very creatures it pretends to explain.

Thirteen

Three Refutations

> A theory is refutable, hence scientific, if it is possible to give
> even one conceivable state of affairs incompatible with its
> truth.
>
> Michael T. Ghiselin

The originator of any radical new scientific theory has an obligation to identify future discoveries that would falsify it. I do so in this chapter. However, before describing the three refutations I have devised, I need to address certain research findings that appear to be in conflict with the theory, but which in fact are not. These are nonanimal genes with the same molecular structure as animal oncogenes.

If oncogenes were originally responsible for opportunistic--and beneficial--plant-like growth in primordial cell colonies (as I suggest in Chapter Two and expand upon in Appendix I) then it is likely that genes with the same molecular structure as functioning oncogenes have survived in plants and other nonanimals. In those creatures the genes continue to function as initiators of beneficial growth; they do not cause cancer.

Such genes include the *ras* gene found in yeast that I mentioned in my published papers. Yeast are unicells and cannot possibly get cancer: yeast *ras* genes are *not* functional oncogenes. Both *ras* and *src*, another oncogene, have been found in sponges and hydroids. My theory states that those multicellular organisms are not products of cancer selection, that their *ras* and *src* genes are not functioning oncogenes. Like yeast *ras* genes, those findings do not refute my theory. Possible future discoveries of other *pseudo*-oncogenes in plants or in other nonanimal multicells would also not refute it.

Now for real tests of my theory, findings that would falsify it. Unlike the neo-Darwinists, and as further indication of my theory's superior power, I need not concoct fantastic and obviously non-existent features. My proposed refutations are far-fetched only if the old theory is wrong and my proposed radical replacement for it is right.

The first falsifying test would attack my assertion that all selected anti-oncogenes are also proreplicative:

Refutation One. Find a morphophysiological character in any prereproductive animal that would putatively protect it from cancer but that would not *also* enhance precise execution of the development program.*

Those who carefully consider the problem will see that such a characteristic is ruled out with tautological certainty by the fact that cancer starts with imperfection in a dividing cell's DNA. They would expend no effort looking for something that cannot exist. However, anyone who thinks that I have erred or that I overestimate the importance of logic in biology is heartily encouraged to attempt this refutation.

The second refutation stems from my assertion, in complete disagreement with neo-Darwinism, that the two distinct groups of multicells (animals and nonanimals) had entirely different mechanistic histories. Animals were products of cancer selection; nonanimals were not. The second falsifying test would be to initiate cancer in a nonanimal.

Refutation Two. Using ultra violet light at intensity levels and duration times found in nature, initiate lethal cancer in a plant

---

*Habitat-selection defenses that provide near-perfect whole-body shelter might, in time, weaken development efficiency; complete absence of cancer might cause relinquishment of somatically-expressed anti-oncogenes. As already explained, such defenses account for the regression of tapeworms to a near-colonial state. For those reasons such defenses are not refutations.

or in another nonanimal multicell, such as a jellyfish, that regularly exposes unshielded and unpigmented somatic cells to direct sunlight.

If the results of that test were positive it would indirectly refute my assertion that sunlight-induced cancer selection was responsible for the evolution of complexity above the tissue level of organization. My final challenge to would-be falsifiers is the most powerful. This one attacks, directly and simply, my central claim that the evolution of complex animals was driven by cancer selection.

Refutation Three.   Find a complex animal with no obvious cancer defenses.

According to Popper, a new theory exhibits strength if it *prohibits* phenomena that were not prohibited by the old theory. Neo-Darwinism does not forbid physical cancer defenses that are not also development-enhancers. My theory prohibits them (Refutation One). Nor does the old theory prohibit, as mine does, sunlight-induced cancer in plants (Refutation Two). And the old theory doesn't prohibit an animal possessing an ensemble of specific properties, all of which exist *separately* in many modern multicells. But my theory again flatly forbids what neo-Darwinism permits. It says that certain animals simply *cannot* exist. Find *one* of these proscribed creatures and my theory is falsified.

My forbidden animal is not narrowly defined. It could be marine, terrestrial or amphibian. It could have any shape. It could subsist on any imaginable diet. It could be predator or prey. It could breed, raise its young, move about and generally behave in any way imaginable. But it must possess, or it must not possess, the following specific characteristics:

It must:

Weigh at least one kilogram at maturity.

Remain a juvenile for at least six months.

Be made up of at least thirty different kinds of cells.

Have a central nervous system including a brain.

Possess a circulatory system complete with a heart.

But it must not:

Have noncellular outer covering, or pigmentation in its exposed cells. (Its external flesh must be soft, hairless and transparent.)

Array masses of postmitotic cells so that they shield dividing cells from solar radiation.

Possess a means of destroying cancer cells similar to the vertebrates' immune system.

Severely limit mitosis; no eutely or reliance on postmitotic cells throughout the soma is permitted.

Finally, to eliminate habitat selection as a cancer defense--

This creature must:

Bask in direct sunlight for at least eight hours a day both as a juvenile and as an adult.

Now that is an *undemanding* list! I require no organs that help other species. I insist on no bizarre luminous spots on fishes' tails. Unlike those pseudo-refutations of neo-Darwinism, all of my characters actually exist. In abundance. In extant multicells. It ought to be easy to find a single organism with that combination of features. A good-sized transparent worm or mollusk living on the surface of a body of water would do the trick.

Biologists may object. They might say that opaque outer coverings and skin pigmentation have nothing to do with cancer selection. Those features (so they might claim) prevent sunburn or dehydration.

My response to these anticipated objections? Go tell it to the siphonophores! The Portuguese-Man-Of-War isn't pigmented. It has no noncellular outer shell and no cancer-specific immune system. It, and its unprotected somatic cells, survive very nicely on the surface of the sea unsheltered from fierce tropical sunlight. Sorry, quibbles about sunburn and dehydration are not acceptable. Just find one complex bilaterian that does for a few hours a day what the plants, siphonophores and jellyfish have been doing for hundreds of millions of years.

According to neo-Darwinism, this animal ought to be out there, somewhere. Evolving animal lineages responded to vacant environmental niches, so biologists have told us, the way gases respond to vacuums--they filled them. The old theory implicitly *predicts*--it certainly does not prohibit--the "unexceptional" combination of unexceptional characters I have specified.

Of course, there *is* no such animal. My theory is correct. Neo-Darwinism is wrong. Animal complexity *was* a product of cancer selection and the lineages that produced complexity also accumulated powerful, easily-identifiable defenses against the disease.

Considering that they have had available for contemplation many millions of extant and extinct species, one would expect professional theorists to have long ago identified the absence of "naked" sun-bathing complex animals as a baffling evolutionary mystery. But they seem not to have even noticed that there aren't any.

Fourteen

The Absurd Alternative

> I find your theory unnecessary in that no phenomena require
> explanation by your hypothesis, unsubstantiated in that you
> offer no evidence for the existence of the entities you
> postulate, and unscientific in that it does not appear to be
> falsifiable.
>
> A biologist, in a letter to the author.

Entertaining the possibility that a plausible radical idea that is
widely-supported by evidence may be wrong elevates to serious consideration an alternative, usually hidden, that is far less likely to be
correct. The concealed alternative is monumental--utterly implausi-
ble--*coincidence*.

Perhaps the best way to approach that greatly under-appre-
ciated risk--one thoughtlessly assumed in the past by many scientific
naysayers--is to turn once more to the professionals' foolish
rejection of Wegener's theory of continental drift.

In rejecting Wegener's idea, the professionals embraced this
alternative: although they were never part of a single land mass that
later broke up, the continents nonetheless acquired shapes consistent with that scenario. In other words, they came to look as if
they had once been connected as the result of *coincidence*. At the
same time they also classified as coincidence (and it may require
some effort to see this) Wegener's plausible explanation of the
apparent fitness.

The appearance of past attachment was no minor observation.
It was overwhelming. If only the east coast of South America

seemed to fit the west coast of Africa it would not have been so bad. But if you attack a world map with a pair of scissors and manipulate the individual continents as if they were pieces of a jigsaw puzzle you will see that Europe and North Africa fit together, North America fits with Europe (east coast to west coast) and Asia (west coast to east coast). Australia fits with Asia, and Antarctica fits somewhere else.

To imagine that all the fitting-together was caused by an astronomical number of *unrelated* accidents is implausibility raised to the highest power. One has to possess a fundamentalist's *worship* of coincidence to prefer it over Wegener's idea. Coincidence was the *wackiest* and *unlikeliest* of all explanations! But decades after the physicists had accepted Einstein's revolutionary theories about time, energy and matter, virtually all geologists (and paleontologists) proudly and openly embraced that nonsensical proposition.

What ought the geologists have done? Well, if they could not honestly disagree with Wegener's assertion that the continents all *looked* as if they had once been a single mass they should have accepted *as fact* their prior attachment. That was the only choice that avoided the alternative of stupendous coincidence.

The professionals might then have said to the meteorologist:

You have obviously noticed something of great importance that we overlooked. It is simply impossible that the correlations you have pointed out could have resulted from a series of unrelated accidents. However, we don't think you are right about the tides; they lack the power needed to move land masses. Because we don't know of a workable alternative we will "black box" the transport problem for now. But we will no longer teach that the continents have been immobile.

The scientists did nothing of the sort. They went on ridiculing Wegener while embracing a notion that had no more connection with reality than the ancients' conviction that the sun revolved around the earth. When tectonic plates were discovered in the 1960s the idea that the continents had never moved went where it belonged: into the dumpster. But the professors' addle-brained faith in absurd coincidence had survived for decades. That such an irrational notion could flourish at major universities around the world in this century at a time when a plausible alternative was on the table speaks volumes about the incredibly poor judgment of those who presumed they were intellectually superior to Wegener and thus capable of evaluating his brilliant insight.

The same potential for embracing absurdity confronts any scientist faced with a radical new proposal, including the one I present in this book. There are no mathematical formulations to express it, but it is clear to me (and ought to be to anyone who thinks about it) that all radical scientific proposals have a peculiar, --*dangerous*, really--property. Simply put, *because* of their radical-ness, they are *more* likely to be correct than less daring ideas, providing they do not conflict with the facts.

Just consider what disagreement with my theory requires the scoffer to assume:

1. Someone *untrained in science* develops a new theory of evolution that:

2. Possesses strong internal logic, *and*

3. Anticipated the finding of cellular oncogenes, *and*

4. Anticipated the finding of cellular anti-oncogenes, *and*

5. Is consistent with a wealth of biological and medical facts, *and*

6. Satisfyingly explains the unbroken chain of precise replication in millions of lineages, *and*

7. Logically explains the origin of phenotypic uniformity, bilateralism, senescence, sleep, sunlight avoidance, weak regeneration, the small size of insects, other long-standing evolutionary puzzles, *and*

9. Meets all the criteria, including falsifiability, demanded of a scientific theory, *but*

10. *It is completely wrong!*

If a scoffer cannot mount a serious argument with statements

1 through 9 but then denies the theory's validity, he is piling coincidence on top of coincidence; he is constructing a ziggurat of
absurdities.

Despite the obvious irrationality of judging my theory to be
wrong, however, I long ago concluded that the pose of skepticism
toward new ideas has a beguiling, ego-inflating quality that many
intellectual mediocrities find too appealing to resist. Those who
adopt it can, after all, feel superior to nearly everyone else. Not
only is the new theory's originator rendered inferior to the skeptic,
but so is anyone who may have been persuaded by his arguments.
All are rendered inferior to the naysayer, at least in the kingdom
of his own small mind.*

But so-called skeptics in the biology community who scorn my
theory should know that their's is a course fraught with peril. Like
those geologists who spurned Wegener's brilliant insight, they risk
nothing less than the ridicule and contempt of future generations.
For they can deny the validity of my idea only by congregating at
that shabbiest of intellectual shrines; they can reject it only by grov-
eling before the Altar of Absurd Coincidence. And as the history
of science demonstrates, coincidence-worship is practiced only by
fools.

---

*As an example of the genre, consider the quotation at the beginning of this
chapter. Those contemptuous words were written by a professor of evolutionary
biology at a major university who included them in a letter rejecting my theory
on behalf of a journal he then edited. This scientist went on to promise, grandly,
an apology if another journal later published my work. His word apparently has
about as much value as his scientific judgment; I never heard from him again.

Fifteen

Reconciliations

Cancer is not a biological entity.

Julian Huxley

To develop this theory I had to revise radically two reigning perceptions: (1) that natural selection is a *sufficient* explanation for the evolution of complex animals and (2) that cancer is a disease. In this chapter I reconcile the new theory with those two old ideas.

The easier of the two reconcilements is with the conception of cancer as a disease. Cancer *is* a disease. It is, however, *also* an ancient biological function of awesome evolutionary significance.

The failure of scientists to see this dual nature of cancer greatly influenced the design of research programs undertaken to understand it. If scientists had been more cautious and did not rule out the possibility that it was also a biological function they might have progressed more quickly to the mutational-oncogenic initiation concept on which I built my theory. They seem to have reached that point--it is now the consensus--but they got there through the plodding, time-consuming, induction-intensive efforts of thousands of researchers.

I came to the same conclusion very quickly, using information known to cancer specialists for decades. The correct consensus may have been reached much sooner if an influential scientist (someone whose ideas would not have been ignored by cancer re-

(someone whose ideas would not have been ignored by cancer researchers) had developed and published this theory decades earlier.*
Having said that, however, I must acknowledge that techniques for
evaluating proposals about the molecular basis of cancer have been
developed only recently.

In retrospect, it seems to me that while twentieth-century researchers were amassing evidence seeming to support the view that plants
and animals were products of the same mechanisms, some evolutionists ought to have expressed discomfort with the fundamental
implausibility of that proposal. As far as I can tell, no one did.
But that great failure notwithstanding, the surviving value of Darwin
and Wallace's original idea--natural selection--and the theory built
on it cannot be denied.

Neo-Darwinism is a sufficient explanation--and this is no small
accomplishment--for the evolution of plants and plant-like multicells. And no organisms, including the far more complex animals,
could have evolved without its mechanisms, especially natural
selection. But the old theory of evolution never was sufficient for
animal evolution. It must be replaced. A new comprehensive
theory of evolution, one that assigns the central transforming role
to cancer selection would be sufficient. It explains both the origin
and the evolution of gene pools capable of producing animals of
great complexity. But cancer selection alone could not have produced *adaptive* complexity. Animal lineages faced other threats
besides cancer and natural selection ensured that only adaptive
organisms survived.

The conclusion is clear. Evolutionary theory must include both
cancer selection--which explains the existence of complexity--and
natural selection--which explains much of the specific *form* of the

---

*An opportunity was missed in the 1950s when a group of cancer specialists
asked the late Julian Huxley, one of the architects of neo-Darwinism, to consider
the disease. They apparently hoped that the famous evolutionist might shed some
light on the great medical puzzle. Huxley accepted the challenge, but, as seen
from the quotation at the beginning of this chapter, he missed a great opportunity.

complexity. I think it important, however, that the relationship between those two types of selection be understood.

I find it helpful to compare the construction of a comprehensive theory of evolution to the erection of a modern skyscraper. My idea, evolution by means of cancer selection, provides the foundation and the steel framework. It gives the structure solid exterior walls and firm, weight-bearing interior floors. It provides a leak-proof roof. It transforms the raw materials into a capacious and solid structure that can withstand the elements, even powerful earthquakes and hurricane-force winds.

With cancer selection recognized as central to animal evolution, the new theory of evolution explains what the old theory could not. It explains the origin of animals and the characteristics that distinguish them from plants. It illuminates the evolution of development, the mechanisms of evolutionary transformation and the function of cancer. And it does so legitimately. It explains through the power of its logic and with only those raw materials permitted to any evolutionary theorist: observable biological mechanisms.

A skyscraper with its foundation firmly fastened to bedrock, heavy steel beams skillfully bolted together, solid floors, walls and a roof is an impressive accomplishment. But it is not of much practical use. It might overwhelm everyone with its size, power and grace, but no one could inhabit it. Much must be done to complete it. Elevators, heat, ventilation, lights, plumbing, partitions, paint, furniture--all the accouterments needed in a functioning building--must be installed. Much costly material and skillfully executed craftsmanship would be needed to convert the structure into a building that people could use.

So it is with a theory of evolution that ignored natural selection and adaptation. Cancer selection gives evolutionary theory the structure it never had. It identifies the biological force that transformed relatively simple animals into extraordinarily complex ones. But it cannot explain--by itself--why animals have all the *particular* features they possess. It explains phenotypic uniformity, the origin of organ-building capability and the transformation of development processes. It can even explain the original function of certain morphological features such as primitive light detectors, opaque external coverings and the human brain. But it cannot explain why bats use echolocation to navigate in the dark or why mammals have an inner ear. It says nothing about altruistic behavior. Yes, cancer selection explains animal complexity, but it doesn't explain *specific* characters and behavior patterns.

The old theory of evolution by means of natural selection has already given us adequate explanations of the adaptive function of many essential animal characteristics.   Those are the interior equipment and furnishings of a comprehensive theory. For that, Darwin, Wallace and those who amplified and clarified their idea of natural selection deserve our continuing appreciation.

But just as it is more sensible to construct the foundation, framework, walls, roof and floors of a building before turning one's attention to the furnishing of its rooms, so it would have made more sense to explain first *how* animals came to exist and *how* they (and their development processes) were transformed before explaining *why* particular products of evolution came to exist. But that's what happened.

Future historians of science may be intrigued or even amused by the fact that many, perhaps most, minor aspects of evolutionary theory were refined long before a solid underlying theory existed. Some of the more percipient contemporary evolutionists might find it embarrassing. But there is nothing anyone can do about it. Like evolution itself, that sequence of events is now a matter of historical fact.

Sixteen

Who Shall Judge?

> Ignorance is preferable to error; and he is less remote from
> the truth who believes nothing, than he who believes what is
> wrong.
>
> Thomas Jefferson

> Two...comrades came to see him about it. One of them had
> worked in the security services and the other for the World
> Trade Union Federation, and both said that biology was a
> science of concern to the party and that there was no place
> on the Soviet-Bulgarian commission for nonparty specialists.
>
> Michael Voslensky

No matter what its intrinsic value, a new theory must attract
supporters if it is to have any influence. That demands evaluation;
prospective backers must decide if they agree with it. But who is
to judge? Who is to decide whether my theory correctly identifies
the single most important factor in animal evolution, or--it's the
only alternative--that it is worthless?

I have strongly-held opinions about who are the best--and the
worst--candidates for evaluation. I'll start with those whom I
consider the least qualified:

**Scientists whose conception of science excludes the hypothe-
tico-deductive method I have used.** My approach is the only
appropriate method for solving problems in historical evolution.
Anyone who says that it is not, or who claims that other
methods are superior, is simply identifying himself as incompetent.

Biologists who turn a blind eye to the structural weaknesses of neo-Darwinism. I do not expect anyone to address *all* its failings but any biologist who does not acknowledge that the old theory cannot explain the persistence of precise development in transformed animal lineages is not serious.

Marxist biologists. I discovered, not long ago, that the followers of that nineteenth century crackpot, Karl Marx, have considerable influence in theoretical biology, particularly in the United States and Great Britain. This is not the place to examine this bizarre situation in any detail, but to give some idea of how weirdly pervasive is the influence of that third-rate thinker in one of our major sciences I cite the following flagrant examples.

Stephen Jay Gould, current President of the Society for the Study of Evolution, Harvard professor and author of many popular books on evolution, seemed to think it important to proclaim in an evolutionary journal--of all places!--that he "learned his Marxism, literally at his Daddy's knee."

Two of Gould's Harvard colleagues [Levins and Lewotin] dedicated a *biology* book to Karl's fellow genius, Friedrich Engels. They did that, not in the 1930s, but in the mid-1980s.

Scientists who contradict Karl's doctrines are routinely savaged by Marxist science-polluters. Favorite targets are biologists who suggest that inheritance may have some influence on human behavior. Apparently Karl proclaimed otherwise. Fortunately for today's heretics, the apparachiks of today's People's Republic of Biology don't kill those who disagree with them. Forced to conform to the standards that prevail in the Western countries in which they and their families (Marxist Daddies included) choose to live, they lack the power of Stalin's favorite biologist, Grigor Lysenko, who routinely arranged for the extermination of Russian scientists who disagreed with the Marxist doctrine that Mendelian genetics was an imperialist plot.*

I don't know (or care) what these cultists think about my
theory, but I caution others not to trust anyone whose commit-
ment to a politico-ideological agenda preclude objective analysis
of ideas, even those concerning the transformation of worm-
genes to elephant-genes.

Biologists who have demonstrated inability to think straight.
I have noted the mushy-mindedness of individual biologists on
particular subjects as well as the collective inability of professional evolutionists to even identify problems deserving intense
theoretical attention.

I won't try to summarize what they *do* consider important but
I think it instructive to report that while at the 1982 annual
meeting of the American Association for the Advancement of
Science, I listened, dumbfounded, to a lecture by the hyper-
active Gould in which he proclaimed before a large audience
(gathered, no less, to honor the centenary of Charles Darwin's
death) theoretical significance for functionless pinhole-sized
holes in the shells of snails found on a Caribbean island.**

---

*Anyone interested in learning more about the bully-boy tactics used by biology's
contemporary Marxies should read the appropriate chapters of the books of Bernard D. Davis and John L. Casti listed in the bibliography.

**To illustrate the theme of his lecture, which was that not everything is adaptive,
Gould showed a slide photograph of the ceiling of a Gothic cathedral. There
were holes in the ceiling. He explained that the holes do not serve any purpose
but exist only because medieval builders could not construct cathedrals without
leaving little openings in the roof. Then he flashed a picture of a snail's shell
with tiny holes in its side, explained that they too were a byproduct of construction and, making a stunning deductive leap, told the audience that because there
are functionless holes in gothic roofs and functionless holes in snails' shells, the
importance of adaptationism--the idea that characters of animals were selected
because they served a function--has been overrated. (Adaptations make Marxies
uncomfortable. They open the door to genetic influence on human behavior.)
    Actually, Gould had a much better example located, so to speak, right under
his nose. I defy anyone to describe the adaptive function of the human armpit.
Like those holes in church roofs and snails' shells, armpits are artifacts of the
architecture employed. Nature, in attaching arms to body trunks, simply couldn't
help creating them.
    I hereby grant Professor Gould permission to use the armpit example in his
campaign to stamp out adaptationism.

Gould's inability to distinguish the significant from the insignificant suggests that he may lack the intellectual skills needed to function effectively in evolution. It is perhaps relevant that this man, who seems compelled to make odd admissions in print, confessed to readers of the March 29, 1984 issue of *The New York Review of Books*:

> I am hopeless at deductive sequencing...I never scored particularly well on so-called objective tests of intelligence because they stress logical reasoning...

Those self-identified weaknesses may explain why Gould committed the "obvious error in logic" that scientist-reviewer James Trefil detected [in the October 22, 1989 issue of *The Washington Post Book World*] at the core of his recent popular book on evolution, and why scientist Bernard D. Davis, commenting on another Gould book, said that he failed to analyze "ideas sharply and with logical rigor, as we have a right to expect of a disciplined scientist." In any event, it would seem imprudent to rely on someone who is "hopeless at deductive sequencing" to evaluate a radical new theory in a field dependent upon logical reasoning.

Scientists who have failed to digest important recent developments. In 1984, Anna Riddiford and David Penny, expanding on the idea that adaptations were not of central importance to evolution, grandly proposed that,

> from an operational viewpoint it is better to assume that a particular [morphological] feature has no advantage unless it can be shown that the feature cannot be explained by chance.

Oh really? Three years after the discovery of functioning oncogenes inside all animal cells these two biologists were claiming that nature had routinely introduced animal characters--all made of somatic cells containing lethal cancer triggers--that provided no survival advantage whatsoever!

Now it's entirely possible that Riddiford and Penny were not aware of my theory. But they simply could not have *not* known--for it was widely acclaimed at the time and was to earn a Nobel Prize for co-discoverers Michael Bishop and Harold Varmus--that oncogenes had been found in normal cells. Once the awesomeness of that discovery sunk in how could any serious evolutionary theorist assert that nature had routinely risked the termination of lineages--by adding new or newly differentiated somatic cells all loaded with oncogenes--for entirely *frivolous* reasons? And then, underscoring their incomprehension, claim they were prudent to do so?

Any evolutionist who demonstrates a clear inability to appreciate the significance of the oncogene discoveries--or, for that matter, Ames' earlier findings that carcinogens are mutagens--is asking the rest of us to ignore him or her.

By this time the reader will sense that I have firm control over my admiration for the work of present-day academic theorists. In fact I am convinced that, unlike all other branches of science, evolutionary theory has made less progress in this century than in the last.

In addition to the unique problems created by Marx's destructive zealots*, modern institutional biology has another pervasive and insidious problem. It is dominated by individuals who were rigorously selected for their ability to learn disciplines that are irrelevant to the demanding task of constructing evolutionary theory. The contrary was true of nineteenth century biology. Possessing little precise knowledge of nature's mechanisms, scientific workers in that era were obliged to *think*. And as Philip Appleman observed of Darwin, they thought in terms of "wholes and continuities rather than in discrete parts and rigidities." Forced to work with black boxes, some of those earlier abstract thinkers--Darwin,

---

*Luckily for practitioners in those sciences, Marx apparently offered no insights on nuclear physics, chemistry or astronomy.

Wallace, Mendel and Weismann--achieved immortal results.*

But many modern biological scientists, including those who have entered evolutionary theory, seem to function under the grand illusion that they have access to most of the relevant data. Certainly, judged by the outpourings in their journals and books, they appear inundated by facts, thoroughly immersed in information. But what modern evolutionists may think of as an ocean of fact is merely a droplet, an infinitesimally small portion of the universe that is--or ought to be--the object of their collective intellectual effort: all the evolutionarily significant events that ever occurred on this planet. Complete and profound intellectual acceptance of the utter inaccessibility of all but a minuscule portion of important biological phenomena ought to lead theorists to reject the methods they were taught to revere--all of them appropriate for dealing with accessible phenomena--and to start thinking like the great theorists of the last century: abstractly and conceptually, in "wholes and continuities."

There are, however, no organized means for scientists to learn approaches that are radically different from those they were taught. They must learn--improvise, actually--on their own. Such self-directed reformation of the human intellect is, however, an extremely difficult undertaking and few modern biologists have accomplished it. Instead, the "ocean" of fact (its illusoriness notwithstanding) determines methodology. Most so-called evolutionary theorists adhere to school-taught techniques. They gather and measure data by direct observation--preferably in experiments--and manipulate it with mathematics and modeling. In issue after issue, *Evolution* consists entirely of reports of *accessible* phenomena, *precisely* observed and measured.

This tyrannical domination of data-centered methodology has had a Gresham-like effect on the science. Its gatekeepers admit to the institutions only those who have mastered the techniques they themselves were taught as students. All others are turned away. The technocratic bean-counters have driven out the think-

---

*Some may object to my inclusion of Mendel among the theorists, claiming that he was an experimentalist and a mathematically-oriented one at that. But the good friar brilliantly devised the concepts of particular inheritance controlled by discrete units (genes) and of dominance and recessiveness, and that kind of creative conceptual thinking is the essence of theorizing.

ers.

So pervasive is this data-worship I am convinced that if he were alive today young Charles Darwin could not acquire accreditation in biology from a prestigious university. His difficulties with higher mathematics would surely have led to the counsel that he seek a career outside of science.

It is, of course, only my opinion that young Darwin couldn't hack it, say, in an MIT doctorate program. But it is neither opinion nor speculation to say that emphasizing mathematics discourages non-mathematical problem-solving. Nor, as I have explained, is it speculative to say that mathematics and experimentalism are of limited use in theory-building.

Based on those facts we need to ask if the administrators of institutional biology, under whose superintendence evolutionary theory ought to flourish, have not created a pervasive mismatch of intellects with problems. Do they not deliberately recruit, train and accredit persons who are *especially* deficient in skills needed in evolutionary biology? Do they not *systematically* exclude future Darwins? Does this explain why it takes a rank outsider like me to see that neo-Darwinism, which was concocted and promoted mainly by experimentalists and mathematically-oriented geneticists, doesn't even come close to achieving its proclaimed central goal?

And if professional biologists are demonstrably unqualified to build evolutionary theory, does it make any sense for intelligent persons who were not rigidly indoctrinated in population genetics, paleontology and other narrow specialties, all of them marginally relevant to evolutionary theory,* to defer to experts in those fields when faced with a problem for which they have *not* been trained: the evaluation of a new general theory of evolution?

Perhaps the question has already occurred to Ernst Mayr, who observed in 1985:

---

*Although some may find this statement impertinent, I developed and explained my theory without reference to matters as fundamental to genetics as dominance and recessiveness, the Hardy-Weinberg law or the structure of DNA. As for fossils, all I needed was a general idea of trends they outline in many animal lineages: sunlight avoidance to sunlight tolerance, simple organisms to complex organisms, marine habitation to terrestrial habitation and the like. It may also be of interest to know that I completed all my basic theoretical work using only information gleaned from an encyclopedia.

> Functional biology...is not too far removed from the
> physical sciences, and is quite congenial with their methods
> [however] *evolutionary biology, with its interest in historical
> processes, is in some respects as closely allied to the humani-
> ties as it is to the exact sciences.* [Emphasis added.]

The idea that scientists are not best qualified to judge new evolutionary theories has important historical precedence.

Visitors to the British Museum of Natural History are greeted
by a greater-than-life-size statue of Sir Richard Owen, who, until
Darwin published his bombshell book, was the most famous nineteenth century English biologist (and the force behind the founding
of the great museum.) But when I entered the museum a few
years ago and gazed up at Owen's likeness I thought the sculptor
ought to have placed a dunce cap on the founder's bald head. For
whenever I think of Owen I don't recall that he founded a great
museum or that he was once the world's leading anatomist. No, I
associate him with that gravest of all scientific errors, the rejection
of truth. For, despite its acceptance by many other biologists,
Owen spurned Darwin's idea of common descent. A shallow, overly ambitious man who was intensely jealous of Darwin's sudden
fame, Owen threw his lot in with the creationists, denounced the
idea of evolution and in so doing immortalized himself as a scientific fool.

Louis Agassiz, another giant of nineteenth century biology, (he
left his name on the door of Harvard's Museum of Natural History), is also remembered as a dunce. He too wagered his immortal scientific soul (or at least his place in science history) against
evolution and on creationism.

Prominent Victorian scientists in other fields also rejected the
idea of evolution, including Lord Kelvin, the best-known physicist
of the era and Sir John Herschel, the distinguished astronomer.

But scientists were not the only readers of *Origin of Species*.
Many thousands of literate lay people read it and talked about it.
They formed opinions of Darwin's revolutionary idea and vigorously
discussed it at the dinner table and in their clubs and pubs. Although many readers were constrained by their religious teachings
from accepting the truth of evolution, others ignored Owen, Kelvin,
Agassiz and other professional rejecters and, evaluating Darwin's

arguments on their own, became convinced evolutionists.* They simply *out-thought* many of the "great" scientific minds of the day. Their amateur judgment of biology's most important idea--that modern organisms descended from common ancestors--was superior to that of the highly-trained scientists who did not let go of the monumentally wrong notion that species were immutable, that God created all life forms *de novo*, that evolution did not happen.

And if many nineteenth century lay readers were smart enough to recognize the truth of evolution while many professionals were disastrously wrong to reject it, why should present-day generalists hesitate to evaluate *my* theory?

Although I have reason to believe that all prominent present-day professional evolutionists read my theory when it was published in the *Journal of Theoretical Biology,* not one of them stepped forward to support it--or to denounce it. Now, with publication of this book, they will all have a second chance. But I am convinced that, even if they could overcome the comprehension problem, most could not acknowledge that an idea developed by an amateur who has had the bad grace to publish it in a book written for the general public warrants serious consideration, let alone acceptance. So concerned are these masters of minutia with their careers and the maintenance of communal decorum, so lacking are they in the courage needed to support a truly radical idea that they will all probably spend what remains of their professional lives locked in the arms of that decrepit old fraud, neo-Darwinism.

But I wrote this book not for scientists but for the general public and I did so with two objectives in mind. First of all, I reject the proposition that only certain employees of universities and

---

*Gertrude Himmelfarb retails two relevant anecdotes. Prior to printing Darwin's book, his publisher, John Murray, asked George Pollock--a lawyer who was not even an amateur scientist--for his opinion. Pollock told Murray it "was probably beyond the comprehension of any scientific man then living," but urged him to double the size of the initial printing because it was "brilliant". Later, Thomas Huxley, Darwin's most enthusiastic scientific supporter, taught a night school course on Darwinism to a group of working men. As he flippantly informed his wife, his students were soon convinced "they are monkeys." It seems that even some of the more modestly educated citizens were miles ahead of the professional naysayers.

museums may construct or evaluate evolutionary theories. Perhaps the biologists *ought* to be competent evolutionists, but if they have failed in their solemn responsibility to interpret facts in such a way that we can better understand our origins--if they are incompetent theorists--then society has every right to pay attention to whomever *has* managed to get it right.

The conclusion, reader, is obvious. The professionals know nothing of relevance that you don't know. *You* can judge whether my theory, or the old theory, is correct.

The professionals will pretend not to care what the public thinks. They see themselves as members of a priesthood and priests are unlikely to admit concern about what the "laity" might think of matters over which they claim sacerdotal rule. But that pomposity can blind and impair scientists, for among the public are people whom they ought to worry about a great deal: the young men and young women who are, or soon will be, sitting in their classrooms and who will eventually replace them. And to do that--to convince *future* biologists that my theory is correct--is the second reason why I wrote this book. For, as the theoretical physicist Max Planck observed,

> An important scientific innovation rarely makes its way by gradually winning over and converting its opponents. What does happen is that its opponents gradually die out and that the growing generation is familiarized with the idea from the beginning.

By going over the heads of the biology teachers I will capture the unbiased and fertile minds of the next generation of evolutionists. And that will be the end of the old theory.

As for that small number of individuals who have dominated published evolutionary biology in recent years, I fear that, like Richard Owen and Louis Agassiz before them, they will be remembered best not for those works to which they have affixed their names, but for their reactions, whether expressed or not, to an idea whose provenance disconcerts them and whose scientific legitimacy, power and truth apparently lie beyond their competency to assess.

Appendix I

The Origin of Cancer

In the fall of 1983, an editor at *New Scientist*, the British weekly that was about to publish a summary of my theory, asked me how oncogenes happened to be in every animal cell. After thinking about her question for a few minutes I responded, lamely, "I don't know."

The question had simply never occurred to me. My radical restructure of evolutionary theory *began* with oncogenes and cancer selection in place. My job as a theorist, so I then thought, was to explain what cancer selection did, not how it came to exist.

I now see that if I don't offer a plausible scenario for its origin I leave myself open to the same criticism I level at neo-Darwinists throughout this book: failure to confront a difficult theoretical problem.

The difficulty arises from the collision of evolutionary logic with biologic reality. Sound evolutionary thinking tells us that selection was the driving force of evolution long before oncogenes came to exist. The same discipline tells us nature never selected genes because of their ability to kill juveniles routinely. But oncogenes--and cancer--exist. They kill juveniles. The fact that oncogenes came to exist 800 million years ago is no reason to duck the problem they--and the logic they defy--create. A comprehensive theory of evolution should address the problem and solve it.

This then is my brief account of how cancer originated.

Cosmologists are convinced that the universe began with a Big Bang, a single cataclysmic event that created all the atoms in the universe. Biological evolution, however, was a far more complex process. It was the result not of a single calamitous event, but of a *series* of Big Bangs, a succession of irreversible occurrences that changed forever the course of life on Earth. Although other events, such as the origin of oxygen, were essential to the formation of existing life forms, I think the great events that occurred *within* animal lineages, or their ancestral line, can be reduced to five. These were the Big Bangs of animal evolution:

> **Origin of life.**
> **Appearance of one-celled organisms.**
> **Origin of sexual reproduction.**
> **Appearance of multicells.**
> **Origin of cancer.**

The sequence of the last three events is important. Cancer's origin followed the origin of sexually reproducing multicells and it resulted from the interaction of three factors that were already in place in the life of even the most primitive of sexually-reproducing multicells. The first two of these crucial elements were two different classes of genes of multicells. The third element was the environment.

In order to explain the two types of genes involved in cancer's origin I must first explain what it is that genes normally do inside cells.

Any gene located in a particular cell does one of three things. (1) It creates proteins. (2) It regulates other genes. Or (3) it does nothing. Genes that create proteins or regulate other genes inside a cell are said to be *active*; those that do nothing are *inactive*. It is also important to remember that virtually all cells of a multicell contain copies of *all* the genes the organism inherited from its parents.

One type of gene that was *inactive* in *most* cells of the earliest multicells (as well as in present-day organisms) were those that orchestrated events leading to formation of the organism's gametes--the sperm and ova that accomplished sexual reproduction when the cell colony reached the adult stage. Those genes, which I will call *s-genes*, remained inactive as long as the cell colonies were juveniles incapable of sexual reproduction.

The second type of genes were *active* in *most* somatic cells *most* of the time. These genes compelled cells to proliferate through mitosis, an activity obviously essential to the development of all multicells. I call genes that encouraged mitotic growth *g-genes*.

The certain existence of those two types of genes in the earliest multicells flows from three simple presumptions: (1) a full complement of genes was in each cell, (2) the organisms sexually reproduced and (3) cell division mechanisms were similar to those in modern multicells. Those modest assumptions assure us that *s-genes* were not active in the first cells cloned from the zygote. If they had been activated immediately following the initial cell division a multicell would not have come into existence.

It is also self-evident that *g-genes* were expressed in most cells in growing colonies. Functioning *g-genes* were needed to form a mass of cells large enough to sustain organismic life. And natural selection would have favored *vigorous* mitotic growth. From observations of modern plants (all of which descended from primordial cell colonies), it is not difficult to imagine many situations in which *g-genes* rescued early multicells through aggressive, opportunistic cell division. If elements essential to the cell colony's survival (sources of moisture or nutrients) were not available in the immediate area, *g-genes* could save the organism by directing the formation of new cells that would search, by aggressive mitotic growth, in adjacent space for the life-sustaining elements. If an external force, such as storm-tossed seas, damaged the cell colony, the *g-genes* could salvage the entire organism by building new cells to replace those lost in the trauma. The ability to generate (and to regenerate) new cells in response to the immediate needs of the colony (phenomena observable in plants) came to exist because they enabled colonies to survive.

But at some point in every multicell's life it had to divert energy from the production of somatic cells to the formation of gametes. The *s-genes* had to become active. And in order to accomplish their primary task of producing gametes, some *s-genes* had to deactivate the *g-genes*. We can see signs of that process, of *s-genes* taking over from *g-genes*, in many modern plants. At the end of the growing season many plants stop producing leaves, branches and roots and form flowers, or buds containing flowers that will blossom the following spring.

Now let's bring the environment into the picture. Although no one knows in what sort of habitat the first multicells originated, it is again useful to make a tautological statement: they came to exist in environments that were favorable to their origin. Simple observation tells us that origins more likely occurred in warm, moist habitats than in cold or dry situations. (Biologists agree that the earliest multicells lived in aquatic habitats.) With an extended, perhaps year-long, growing season (such as found in tropical seas), the original multicells would have had lots of time for *g-genes* to establish healthy organisms before the *s-genes* intervened to arrange production of gametes.

I now propose that among the multicell lineages whose descendants have survived to the present there was one whose cozy, lush environment disappeared. Its habitat changed from one that coddled the multicells to one that placed them *in extremis*.* The change may have been introduced slowly, over many thousands of years, or perhaps it was more sudden. The cause of the environment's degradation might have been the encroachment of a glacier, a sudden shift of tectonic plates or some other cataclysmic event. Or the organisms might simply have been transported by sea currents to a harsher place, perhaps closer to a polar ice cap. Whatever the circumstances, that population found itself in an environ-

---

*Although the idea that this lineage originated in a less harsh environment is more plausible, it is possible that it never enjoyed a habitat of abundance. It doesn't matter. The sequence of events I am about to describe would have occurred.

ment that did not permit *g-genes* and *s-genes* to maintain the (more or less) leisurely *modus operandi* permitted in balmier climates.

Time was now the crucial factor. The new environment gave the DNA less *time*--perhaps only a few weeks--to create a new generation of organisms. The multicells had less *time* to develop, to breed and to create the viable fertilized eggs that would become the next generation. Perhaps there were very short summers followed by long, harsh winters. Perhaps the organisms lived in isolated pools formed during a brief rainy season that dried up a few weeks later. Whatever the specific circumstances, the genes in that lineage were in a tough environment, one that precluded a leisurely construction of multicells.

S-genes had to seize command more quickly. They had to intervene earlier in each organism's life to ensure that a mature multicell was completed before the brief growing season ended. *G-genes* would still function during the first surge of cell creation, but if that lineage was to survive, *g-genes* had to be deactivated quickly to permit gamete formation before the growing season ended. The alternative was extinction.

The *s-genes* prevailed. Natural selection saw to it. When an individual's *g-genes* were not suppressed soon enough it never reached the reproductive stage before the growing season ended. It's genes were exterminated. Whenever that happened *s-genes* automatically increased their relative power in the surviving pool of DNA. And whenever organisms functioning under the increasingly strict regime of the *s-genes* (turn off *g-genes*! make gametes! *fast!*) bred and left offspring, the *s-genes'* increasingly powerful control over the population was reinforced.

While the extreme selection pressure imposed by the demanding environment was forcing *s-genes* to suppress the *g-genes* early, germ line mutations continued to flow into the lineage. Any mutations that added to the *s-genes'* control over organism construction were selected.

Throughout that period of increasing s-gene power, mistakes continued to occur. Cells were misreplicated; somatic mutations occurred. On at least some of those occasions, *G-genes* would break the fetters imposed by *s-genes* and force a developing organism to grow aggressively, in the old pattern. Once begun (and it could begin in any cell that divided) the rapid growth would not stop. All the descendants of the first wild cell inherited DNA with activated *g-genes*. Although the individual colony might survive an outburst of old-style growth, because that growth could interfere with gamete production the cumulative effect of escape of *g-genes* was increased selection pressure for even more powerful *s-gene* control over the organisms.

Eventually, the degree of somatic order imposed by the *s-genes* reached the stage where escape of *g-genes* in *some* organisms led to its death. And when an even higher level of precision in body construction was reached (due to the accumulation of the controlling *s-genes*) *any* reversion to old-style vegetative growth invariably caused lethal havoc.

And that was it. Biology's final Big Bang had exploded. Animals (for that is what those multicells became) began to die when the strict control over *g-genes* failed inside a single cell. They began to die from the same kind of growth--aggressive and opportunistic--that had benefited their ancestors and that is still beneficial in early animal life and throughout much of the plants' lives. They began to die of cancer.

This explanation for cancer's origin does not conflict with the imperative that lethal genes were never selected. Notice that I do not suggest that dangerous oncogenes (which is what the *g-genes* became) and the potential to die of cancer were *selected*. The transformation of essential *g-genes* (no multicell could exist without them) to potentially lethal oncogenes was the by-product of the selection, imposed by the harsh environment (and consistent with evolutionary logic), of *s-genes* capable of deactivating *g-genes* early in organismic life. (Those *s-genes*, of course, are now anti-oncogenes).

This explanation is also consistent with the relevant modern cancer evidence summarized in Chapter 11.

Appendix II

The *Journal of Theoretical Biology* Papers

The following appeared as a Letter to the Editor in the April 21, 1983 edition of the *Journal of Theoretical Biology*, volume 101, number 4, pages 657-659.

Cancer and Evolution: Synthesis

Cancer has certain characteristics that lead me to conclude that it functions as an "enforcer" of the genetic program and, as such, played a major role in the origin and evolution of the Bilateria. In this theoretical conception of the cancer process, each juvenile specimen capable of getting the disease can be viewed as a "black box" in which the input of a carcinogen results in the output of cancer. Ames' correlation (Ames et al., 1973) permits the substitution of "mutagen" for carcinogen, and cancer's lethality suggests that the output can be labeled "genetic death." Although the precise cellular mechanisms involved in carcinogenesis are not considered here, it is assumed that within a target pre-mitotic cell the following sequence takes place: (a) the mutagen causes a mutational event and (b) oncogenes (Bishop, 1982) initiate transformation to the cancerous state following mitosis. It follows from this sequence that virtually all selected defenses against cancer would have enhanced the ability of the genomes to create organisms in which the genetic program is expressed with great fidelity in all somatic cells.

If, as I believe to be the case, all Bilateria have oncogenes in every cell and descended from species that endured significant losses to mutagen-induced cancer, then all Bilaterian specimens possess an extensive array of characters that function both as cancer defenses and replication enhancers. There would seem to be five basic means by which selected characters would carry out these two functions: (1) minimize pre-reproductive mitosis, (2) avoid exposure to mutagens, (3) provide morphophysiological protection against mutagens, (4) repair mutagen-induced damage before transformation, and (5) destroy or contain transformed cells. Examples of such mechanisms and comments on their replication-enhancive character follow.

Pre-reproductive mitosis avoidance is evident in the strict control over somatic cell production in Bilateria. Compared to plants and many species of tissue-level animals, most Bilaterian specimens lack phenotypic plasticity, regenerate damaged tissue less flamboyantly, undergo diminishment in mitosis following (likely) sexual reproduction, have many fixed post-mitotic cells (neurones and muscle cells) and, in the insects, exhibit polyploidy and polyteny. The possibility of replication errors is decreased when somatic cell production is minimized by these devices.

Minimization of exposure to mutagens is evident in the Bilateria's unique (for multicells) history of sun avoidance. The earliest Bilateria lived either in the sea bottom or on its surface (Valentine, 1973). All of the animals in the latter habitat were equipped with non-cellular outer coverings that protected them from any mutagenic, and carcinogenic, radiation that penetrated the layers of water above. Descendants of those first complex animals all reflect selection of devices that protect against radiation: habitats, including parasitical sites, affording good shelter;

shells, chitinous scales, exoskeletons, feathers, hair and skin pigmentation; small body size, which would have minimized the likelihood of a single lethal, i.e., cancer-inducing, "hit" of radiation; or they were equipped with a "fail safe" immunological system.

Efficacious repair of damaged DNA by repair enzymes functioning in somatic cells would both avoid death from cancer, as I have conceptualized it, and enhance somatic expression of the genetic program.

Complete destruction of malignant cells would eliminate effects of the mutational event that initiated the cancer process. Harshbarger (1968) has suggested that some invertebrates may use autectomy to rid the body of tumors or encapsulation to contain them; metamorphosis, during which diseased larval cells could be discarded, was mentioned by Gateff and Schneiderman (1968) as a factor which might account for low cancer rates in insects. Such disposal or containment of cancer cells would be enhancive of precise replication since these processes, if successful, would help juveniles to survive to reproductive age without being devastated by the effect of a single error in replication.

The evolution of efficient cancer-specific immunological defenses in all vertebrates would have enabled those species to adapt characters, functions, etc.. which might have increased the incidence of cancer initiation. The following all suggest the lowering of first line defenses against cancer in vertebrates: increased mitosis as evidenced by large body size and extended pre-reproductive life, increased exposure to radiation as the result of migration from aquatic to terrestrial habitats, and the elimination, in many mammalian species, of opaque external protection from UV radiation. Bilaterian invertebrates do not have a lymphoid system which, according to Good & Finstad (1968), has as its primary raison d'etre surveillance against malignancy. Unlike animals equipped with such immune systems, the invertebrate germ lines seem not to have produced any large, long-lived terrestrial specimens, and none seem to have shed ancestral radiation-protective shielding to the extent found in some vertebrate species. On the other hand, as noted by Gateff & Schneiderman (1968), experimental data suggest that in the largest group of terrestrial invertebrates, the insects, somatic cells exhibit karyotypic and genetic program stability greatly in excess of that found in vertebrates.

If evolutionary theory is modified to include the assertion that cancer established, about 700-800 million years ago, the imperative that only those Bilaterian genotypes capable of extreme precision in the construction of multi-celled organisms could possible survive to participate in the struggle for existence, and ruthlessly enforced that imperative ever since, then evolutionary theory is strengthened. It would offer, as it does not now, a mechanistic explanation for a generally ignored, but nonetheless perplexing problem: why, if they had access to the same mechanisms as the Bilateria, did the germ lines of plants, Porifera and Coelenterates not create multicells with complex vital organs? Or, conversely, if tissue-level multicells were sufficiently adapted to ensure the survival of their germ lines for hundreds of millions of years, why do organisms of so much greater complexity exist in such abundance in the Bilateria?

D.A.Buyske, W.Bock and R.Milkman commented, helpfully, on earlier drafts. J.C. Harshbarger, R.Dawkins, J.E.Trosko, A. Zeitlin and R.G.Brenner gave me helpful advice or information.

James Graham

[Deletion]

(Received 1 June 1981, and in revised form 23 October 1982)

REFERENCES

Ames, B.N., Dunston, W.E., Yamasaki, E. & Lee, F.D. (1973). Proc. natn. Acad.

Sci. U.S.A. **70**, 2281.
Bishop, J.M. (1982). Sci. Am. **246**, 81.
Gateff, E. & Schneiderman, H.A. (1968). In: Neoplasms and Related Disorders in Invertebrates and Lower Vertebrate Animals (Dawe, C.J. & Harshbarger, J.C. eds), Monograph 31, p.365. Washington: National Cancer Institute.
Good, R.A. & Finstad, J. (1968). In: Neoplasms and Related Disorders in Invertebrates and Lower Vertebrate Animals (Daw, C.J. & Harshbarger, J.C. eds), Monograph 31, p.41. Washington: National Cancer Institute.
Harshbarger, J.C. (1968). In: Neoplasms and Related Disorders in Invertebrates and Lower Vertebrate Animals (Daw, C.J. & Harshbarger, J.C. eds), Monograph 31, p.xi. Washington: National Cancer Institute.
Valentine, J.W. (1978). Sci. Am. **239**, 140.

* * *

The second Letter appeared in the March 21, 1984 edition, volume 107, number 2, on pages 341-343 of the journal:

Cancer and Evolution: Amplification

To amplify the ideas expressed in "Cancer and Evolution: Synthesis" (Graham, 1983) I suggest that all Bilaterian genetic material can be divided into four groups: oncogenes, anti-oncogenes, adaptive pro-oncogenes and those that are cancer neutral.

Genes that are cancer neutral are those whose selection was followed, for whatever reasons, neither by an increase nor a decrease in the incidence of cancer in the organisms equipped with them. These genes have no value in this theory.

In this context "oncogenes" are cellular oncogenes. These are further defined as having the potential for killing the organism in whose genetic program they are present, such deaths being initiated by the occurrence of a mutational event in a single somatic cell. This theory states that oncogenes, thus defined, have been present in every cell of every specimen of every species of the Bilateria that ever existed, and that they have existed nowhere else in nature. Although it has been assumed by many "...that c-onc genes serve some essential purposes in uninfected cells" (Bishop & Varmus, 1982) whether or not oncogenes do indeed have any functions other than to kill organisms is irrelevant to the development of, and the validity of, this theory.

Recent findings by Simon, Kornberg & Bishop (personal communication, 1983) indicating that the src oncogene is present in the genome of Drosophila Melanogaster, and the Shilo & Weinberg (1981) report of Caenorhabditis elegans nematode DNA hybridizing to oncogenes suggest that oncogenes are present in all Bilaterian invertebrates. As for vertebrates, as noted by Bishop (1982), "Of 17 retrovirus oncogenes identified to date, 16 are known to have close relatives in the normal genomes of vertebrate species". Perhaps of equal significance to these molecular findings is Harshbarger's (1980) report of "...the strongest candidate neoplasm yet seen in Platyhelminthes, a phylum at the primitive level of only two germ levels". Viewed in conjunction with Gateff & Schneiderman's (1968) report of lethal and transplantable neoplasia in Drosophila Melanogaster, this pathological finding suggests support for the conclusion that all Bilaterian invertebrates, as well as all vertebrates, have the potential to die of cancer.

Although the reports of DeFeo, Papageorge, Stokes, Temeles & Scolnick (personal communication, 1983) and of Hammond & Bishop (personal communication, 1983) indicate that DNA homologous to ras and fps oncogenes, respectively, is present in yeast, these are not oncogenes as defined since it is assumed that single-celled organisms cannot die of cancer.

Anti-oncogenes are defined as those which were originally selected because they helped to reduce genetic losses to cancer. Because the process leading to such genetic deaths is believed to begin with <u>imprecise</u> replication of the genetic program in a single cell, I conclude that all anti-oncogenes also function as enhancers of <u>precise</u> replication. The report by Yunis (1983) that "High resolution banding techniques...have revealed that malignant cells of most tumors analyzed have characteristic chromosome defects", seems to lend support to the idea that oncogenes are activated in response to mutational events, and therefore, that selected anti-oncogenes would tend to minimize the incidence of these potentially lethal occurrences.

Adaptive pro-oncogenes are those that imparted some survival benefit to the germ line in spite of a likely increase in juvenile deaths from cancer following their selection. Increased somatic complexity, greater body size, extended pre-reproductive life and migration to more mutagenic habitats occurred in so many Bilaterian lineages that they can be confidently judged to have been adaptive. It is, however, most probable that selection of such characters was followed by increases in the incidence of somatic mutational events in juveniles and resulted in increased losses of genetic material to cancer.

The concept of genes that were both adaptive and pro-oncogenic would explain what Mayr (1982) calls transformational, or vertical, evolution in the Bilateria; and it would account for the persistence, at least in some species, of juvenile cancer 700-800 million years after its presumed origin (Graham, 1983). Selection of adaptive pro-oncogenes would have increased the pressure for more effective anti-oncogenes, which, because of their inherent replication enhancive properties, would have enabled the surviving gene pools to create the more complex (or larger or more exposed) animals whose development was by then imbedded in the genetic program. The relative volume of adaptive pro-oncogenes (and anti-onco-genes) selected over time would explain the existence in modern Bilateria of both relatively simple animals and those that are very complex. Those germ lines that created the most complex animals endured the most genetic losses to cancer and vice versa.

This idea is supported by the relative lack of complexity in Bilaterian animals whose ancestors seem not to have ventured from shelters that afford good protection from sunlight and other carcinogenic radiation: earthworms are not as complex as insects, and all bivalves are simpler than the octopus. This pattern would seem to require a mechanistic explanation that is exclusive to the Bilateria, for, although there are no extant or extinct species of large-bodied, relatively simple Bilateria in exposed habitats, the combination of large bodies, relative simplicity and exposure to sunlight is observable in many plants and <u>Coelenter-ates</u>.

I thank two anonymous referees for their comments on the initial draft, the authors cited in the text for granting permission to refer to their unpublished work, G. Ruggerieri, G. Lewis and N. Macbeth for kindly answering questions I posed to them in the early stages of this effort and four anonymous referees who commented on earlier drafts submitted to three journals.

James Graham

[Deletion]

(Received 3 June 1983, and in final form 25 October 1983)

REFERENCES

Bishop, J.M. (1982) <u>Sci. Am.</u> **246**, 81.
Bishop, J.M. & Varmus, H. (1982). In <u>RNA Tumor Viruses</u> (Wiess, R., Tiche, N., Varmus, H. & Coffin, J., eds). p 999. Cold Spring Harbor, New York: Cold Spring Harbor Laboratories.
Gateff, E. & Schneiderman, H.A. (1968). In: <u>Neoplasms and Related Disorders in</u>

Invertebrates and Lower Vertebrate Animals (Daw, C.J. & Harshbarger, J.C. eds), Monograph 31, p.365. Washington: National Cancer Institute.

Graham, J. (1983). J. theor. Biol. 101, 657.

Harshbarger, J.C., (1980). Activities Report, Registry of Tumors in Lower Animals: 1979 Supplement p. 1. Washington: Smithsonian Institution.

Mayr, E. (1982). The Growth of Biological Thought: Diversity, Evolution and Inheritance. Cambridge, Massachusetts: The Belknap Press of Harvard University Press.

Shilo, B-Z. & Weinberg, R.A. (1981). Proc. natn. Acad. Sci. U.S.A. 78, 6789.

Yunis, J.J. (1983). Science 221, 227.

* * *

These papers are reprinted with the kind permission of the copyright owner, Academic Press Inc. (London) Ltd.

Selected Bibliography

Altman, A.A. and Schwartz, A.D. *Malignant Diseases in Infancy, Childhood and Adolescence.* Philadelphia: W. B. Saunders Co., 1978.

Altman, Lawrence K. "Advances in Treatment Change Face of AIDS." *The New York Times,* June 12, 1990.

Ames, B.N., Dunston, W.E., Yamasaki, E. and Lee, F.D. "Carcinogens are Mutagens." *Proceedings of the National Academy of Science. Vol. 53,* June, 1978.

Anders, Fritz. "The Mildred Scheel 1988 Memorial Lecture: A Biologist's View of Human Cancer." *Modern Trends in Human Leukemia VIII.* R.D. Neth, *et al,* (eds.) Berlin: Springer-Verlag, 1989.

-------------, Anders, Annerose, Schartl, Manfred, Gronau, Thomas, Lueke, Wolfgang, Schmidt, Carl-Rudolf, and Barnekow, Angelika. "Oncogenes in Development, Neoplasia, and Evolution." *New Experimental Modalities in the Control of Neoplasia.* Chandra, Prakash, (ed.) New York: Plenum Publishing, 1986.

Angier, Natalie. *Natural Obsessions: The Search for the Oncogene.* Boston: Houghton, Miflin Company, 1988.

Andervont, H.B. and Dunn, T.B. "Occurrence of Tumors in Wild House Mice." *Journal of National Cancer Institute.* 28: 1153-1162, 1962.

Andrew, Warren. "Tumors and Aging." *Monograph 31: Neoplasms and Related Disorders of Invertebrates and Lower Vertebrate Animals.* Dawe, Clyde J. and Harshbarger, John C., (eds.) Bethesda: National Cancer Insitute, 1968.

Appleman, Philip. "Darwin: On Changing the Mind." *Darwin: A Norton Critical Edition.* (Second Edition) Appleman, Philip, (ed.) New York: W.W. Norton & Company, 1979.

Arthur, Wallace. *Mechanisms of Morphological Evolution: A Combined Genetic, Developmental and Ecological Approach.*

Chichester: John Wiley & Sons, 1984.

------------ *A Theory of the Evolution of Development*. Chichester: John Wiley & Sons, 1988.

Ayala, Francisco. "The Mechanisms of Evolution." *Scientific American*, September, 1978.

Bacq, Z.M. & Alexander, P. *Fundamentals of Radiobiology*. New York: Permagon Press, 1961.

Barnckow, Angelika and Mueller, Werner A. "An *src*-related tyrosine kinase activity in the hydroid, *Hydractina*." *Differentiation*. (1986) 33:29-33.

Bonner, John Tyler. *The Evolution of Complexity By Means of Natural Selection*. Princeton: Princeton University Press, 1988.

Braun, Armin C. *The Story of Cancer*. Reading: Addison-Wesley Publishing Company, 1977.

Cairns, John. "Mutational Selection and the Natural History of Cancer." *Nature*. Vol. 255, May 15, 1975.

------------ *Cancer: Science and Society*. San Francisco: W.H. Freeman and Company, 1978.

Casti, John L. *Paradigms Lost: Images of Man in the Mirror of Science*. New York: William Morrow and Company, Inc., 1989.

Ceram, C.W. translated by E.B. Garside and Sophie Wilkins, Gods, Graves, and Scholars. (Second Edition) New York: Vintage Books, 1986.

Conklin, Groff. "Cancer and Environment." *Cancer Biology*. Friedberg, Errol C., (ed.) New York: W.H. Freeman and Company, 1985.

Crossner, Kenneth Alan. "More on the Thymus." *The New York Times*, November 4, 1990.

Dafni, A. and Bernhardt, P. "Pollination of Terrestrial Orchids of Southern Australia and the Mediterranean Region: Systematic, Ecological and Evolutionary Implications." *Evolutionary Biology* Volume 24, Hecht, Max, Wallace, Bruce and MacIntyre, Ross J., (eds.) New York: Plenum Press.

Darwin, Charles. *The Origin of Species by Means of Natural Selection*. (Reprint of First Edition) Middlesex: Penguin Books, 1968.

Davis, Bernard D. *Storm Over Biology: Essays on Science, Sentiment, and Public Policy*. Buffalo: Prometheus Books, 1986.

Dawkins, Richard. *The Selfish Gene*. New York: Oxford University Press, 1977.

------------ *The Extended Phenotype: The Gene as the Unit of Selection*. Oxford: Oxford University Press, 1982.

------------ *The Blind Watchmaker: Why the Evidence of Evolution Reveals a Universe Without Design*. New York: W.W. Norton & Company, 1986.

------------ and Krebs, J.R. "Arms Races Between and Within Species." *Proc. Roy. Soc. London.* 205:489-511, 1979

Dawe, Clyde J. "Phlogeny and Oncogeny." *Monograph 31, Neoplasms and Related Disorders of Invertebrates and Lower Vertebrate Animals.* Dawe, Clyde J. and Harshbarger, John C., (eds.) Bethesda: The National Cancer Institute, 1968.

Devoret, Raymond. "Bacterial Tests for Potential Carcinogens." *Cancer Biology.* Friedberg, Errol C., (ed.) New York: W.H. Freeman and Company, 1985.

Dorst, Jean and Dandelot, Pierre. *A Field Guide to the Larger Mammals of Africa*. London: Collins.

Duellman, William E. and Trueb, Linda. *Biology of Amphibians*. New York: McGraw-Hill Book Company, 1986.

Edey, Maitland A. and Johanson, Donald C. *Blueprints: Solving the Mystery of Evolution*. Boston: Little, Brown and Company, 1989.

Eldridge, Niles. *Time Frames: Rethinking of Darwinian Evolution and the Theory of Punctuated Equilibria*. New York: Simon and Schuster.

Finch, E.C. and Hayflick, L. *Handbook of the Biology of Aging.* New York: Van Nostrand Reinhold, 1977.

Fisher, Lawrence M. "Hopeful of New Cancer Treatments, Companies Team Up to Study a Gene." *The New York Times*, November 29, 1989.

Fishman, William H. and Sell, Stewart, (eds.) *Onco-Developmental Gene Expression*. New York: Academic Press, Inc., 1976.

Folkman, Judah. "The Vascularization of Tumors." *Cancer Biology.* Friedberg, Errol C., (ed.) New York: W.H. Freeman and Company, 1985.

Friedberg, Errol C. *Cancer Biology.* Friedberg, Errol C., (ed.) New York: W.H. Freeman and Company, 1985

Futuyma, Douglas J. *Evolutionary Biology*. Sunderland: Sinauer Associates, Inc., 1979.

Ghiselin, Michael T. *The Triumph of the Darwinian Method.* Chicago: The University of Chicago Press, 1984.

------------ "The Origin of Molluscs in the Light of Molecular

Evidence." *Oxford Surveys in Evolutionary Biology* Volume 5, Harvey, Paul H. and Partridge, Linda, (eds.) Oxford: Oxford University Press, 1988.

Glick, Thomas C., (ed.) *The Comparative Reception of Darwinism.* Austin: University of Texas Press, 1972.

Good, Robert A. "Immunologic Aberrations: The AIDS Defect." *The AIDS Epidemic.* Cahill, Kevin M., (ed.) New York: St. Martin's Press, 1983.

Gould, Stephen Jay. *Ever Since Darwin: Reflections in Natural History.* New York: W.W. Norton & Company, 1977.

------------ "Triumph of a Naturalist." *The New York Review*, March 29, 1984.

------------ "The Meaning of Punctuated Equilibrium and its Role in Validating a Heirarchical Approach to Macroevolution." *Perspectives on Evolution.* Milkman, Roger, (ed.) Sunderland: Sinauer Associates, Inc., 1982.

------------ and Eldridge, Niles. "Punctuated Equilibria: The Tempo and Mode of Evolution Reconsidered." *Paleobiology* 3 (1977).

Goyette, Michele, Petropoulous, Christos J., Shank, Peter R., Fausto, Nelson. "Expression of a Cellular Oncogene During Liver Regeneration." *Science*, February 4, 1983.

Graham, James. "Cancer and Evolution: Synthesis." *Journal of Theoretical Biology*, **101**, 657. April 21, 1983.

------------ "Cancer and Evolution: Amplification." *Journal of Theoretical Biology*, **107**, 341. March 21, 1984.

Gryzimek, B. *Gryzimek's Animal Life Encyclopedia.* New York: Van Nostrand Reinhold, 1974.

Hayflick, Leonard. "Human Cells and Aging." *Scientific American*, March, 1968.

------------ "The Cell Biology of Aging." *Scientific American*, January, 1980.

Headstrom, Richard. *The Weird and the Beautiful.* New York: Cornwall Books, 1984.

Himmelfarb, Gertrude. *Darwin and the Darwinian Revolution.* New York: W.W. Norton & Company, Inc., 1962.

Holton, Gerald. "Einstein's Model." *The American Scholar*, Summer, 1979.

Huff, Clay G. *A Manual Of Medical Parisitology.* Chicago: The University of Chicago Press, 1943.

Huxley, A.F. "How Far Will Darwin Take Us?" *Evolution from*

Molecules to Men. Bendall, D.S., (ed.) Cambridge: Cambridge University Press, 1983.

Huxley, Julian. *Biological Aspects of Cancer.* New York: Harcourt, Brace & Co., 1958.

Hyman, L.H. *The Invertebrates: Platyhelminthes--Rynchocoela.* New York: McGraw Hill, 1951.

------------ *The Invertebrates: Smaller Coelomate Groups.* New York: McGraw Hill, 1959.

Kaestner, A. *Invertebrate Zoology.* New York: Interscience Publishers, 1964.

Kuhn, Thomas S. *The Structure of Scientific Revolutions.* (Second Edition, Enlarged.) Chicago: The University of Chicago Press, 1970.

Lentz, T. L. *The Cell Biology of Hydra.* Amsterdam: North-Holland Publishing Company, 1966.

Levins, Richard and Lewotin, Richard C. *The Dialectical Biologist.* Cambridge: Harvard University Press, 1985.

Lewin, Roger. "Why is Development So Illogical?" *Science,* June 22, 1984.

Lewontin, Richard C. *The Genetic Basis for Evolutionary Change.* New York: Columbia University Press, 1974.

------------ "Adaptation." *Scientific American,* September, 1978.

Lilienfeld, Abraham M., Levin, M.L., Kessler, I.I. *Cancer in the United States.* Cambridge: Harvard University Press, 1972.

Luce, Gay Gaer. *Body Time: Physiological Rhythms and Social Stress.* New York: Pantheon Books, 1971.

Magnus, Knut. *Trends in Cancer Incidence.* Washington: Hemisphere Publishing Corporation, 1982.

Marcus, Erin. "Scientists Pinpoint Gene That Shortens Life Span." *The Washington Post,* August 24, 1990.

Maynard Smith, John. *On Evolution.* Edinburgh: Edinburgh University Press, 1972.

------------ *The Problems of Biology.* Oxford: Oxford University Press, 1986.

Mayr, Ernst. "Evolution." *Scientific American,* September, 1978.

------------ *The Growth of Biological Thought: Diversity, Evolution, and Inheritance.* Cambridge: Harvard University Press, 1982.

------------ *Toward a New Philosophy of Biology: Observations of an Evolutionist.* Cambridge: Harvard University Press, 1988.

------------ "How Biology Differs from the Physical Sciences."

206   *Cancer Selection*

Evolution at the Crossroads: The New Biology and the New Philosophy of Science. Depew, David J. and Weber, Bruce H., (eds.) Cambridge: The MIT Press, 1985.

Medawar, Peter. *Pluto's Republic.* Oxford: Oxford University Press, 1984.

------------ *The Uniqueness of the Individual.* (Second Edition) New York: Dover, 1981.

Mostofi, F.K. and Leestma, J.E. *Pathology.* Sixth Edition, Anderson W.A.D., (ed.) St. Louis: Mosby, 1971.

Nitecki, Matthew H., (ed.) *Evolutionary Progress.* Chicago: Chicago University Press, 1988.

Patterson, Colin. *Evolution.* London: British Museum (Natural History), 1978.

Pereira-Smith, O.M. and Smith, J.R. "Evidence for the Recessive Nature of Cellular Immortality." *Science,* September 2, 1983.

Peters, Esther C., Halas, John C., and McCarty, Harry B. "Calicoblastic Neoplasms in *Acropora palmata,* With a Review of Reports on Anamolies of Growth and Form in Corals." *Journal of the National Cancer Institute,* May, 1986.

Popper, Karl, R. *The Logic of Scientific Discovery.* (2nd Edition) New York: Harper & Row, 1968.

------------ *Objective Knowledge: An Evolutionary Approach.* Oxford: Oxford University Press, 1972.

Radany, William E. "Isolation and Expression of Genes with Homology to the Trysonine Kinase Domain of Viral Oncogenes During Sea Urchin Embryogenesis." *Advances in Gene Technology: The Molecular Biology of Development.* Voellmy, Richard W., *et al,* (eds.) Cambridge: Cambridge University Press, 1987.

Rensberger, Boyce. "What Made Humans Human?" *The New York Times Magazine,* April 8, 1984.

Riddiford, Anna and Penny, David. "The Scientific Status of Modern Evolutionary Theory." *Evolutionary Theory: Paths Into the Future.* Pollard, J.W., (ed.) Chichester: John Wiley & Sons, 1984.

Ridley, Mark. *The Problems of Evolution.* Oxford: Oxford University Press, 1985.

Rose, M.R. "The Evolution of Senescence." *Evolution: Essays in Honour of John Maynard Smith.* Cambridge: Cambridge University Press, 1985.

Ruddon, Raymond W., *Cancer Biology, Second Edition.* New York:

Oxford University Press, 1987.

Ruse, Michael. *Darwinism Defended: A Guide to the Evolution Controversies*. Reading: Addison-Wesley Publishing Company, 1982.

------------ *Taking Darwin Seriously: A Naturalistic Approach to Philosophy*. Oxford: Basil Blackwell, 1986.

Schmidt, Eric. "Scientists Study Fish For Clues to Cancer." *The New York Times*, December 26, 1989.

Shilo, Ben-Zion and Weinberg, Robert A. "DNA sequences homologous to vertebrate oncogenes are conserved in *Drosophila melanogaster.*" *Proceedings of National Academy of Sciences, USA*, November, 1981.

Slatkin, M. "Somatic Mutations as an Evolutionary Force." *Evolution: Essays in Honour of John Maynard Smith*. Cambridge: Cambridge University Press, 1985.

Smith, Anthony. *The Human Pedigree*. New York: J.B. Lippincott Company, 1975.

Stanley, Steven M. *Macroevolution: Pattern and Process*. San Francisco: W.H. Freeman & Co., 1979.

Strehler, Bernard L. "Aging in Coelenterates." *The Biology of Hydra*. Lenhoff, Howard M. and Loomis, W. Farnsworth, (eds.) Miami: University of Miami Press, 1961.

Stevens, William K. "Monarchs' Migration: A Fragile Journey." *The New York Times*, December 4, 1990.

Takahashi, Yasuro. "Growth Hormone Secretion Related to the Sleep and Waking Rhythm." *The Functions of Sleep*. Drucker-Colin, R., Shkurovich, Mario, and Sterman, M.B., (eds.) New York: Academic Press, 1979.

Temin, Howard M. "RNA-Directed DNA Synthesis." *Cancer Biology*. Friedberg, Errol C., (ed.) New York: W.H. Freeman and Company, 1985.

The Economist. "Rooting Out Cancer." September 19, 1981.

------------ "The Delphic Warning of Cancer Genes." February 17, 1990.

Trefil, James. "Survival of the Luckiest." *The Washington Post Book World*. October 22, 1989.

Valentine, James W. "The Evolution of Multicellular Plants and Animals." *Scientific American*, September, 1978.

Voslensky, Michael. *Nomenklatura: The Soviet Ruling Class, An Insider's Report*. Garden City: Doubleday & Company, Inc.,

1984.

Wardle, Robert A. and McLeod, James Archie. *The Zoology of Tapeworms*. Minneapolis: The University of Minneapolis Press, 1952.

Weinberg, Robert A. "A Molecular Basis of Cancer." *Scientific American*, January, 1983.

Whyte, Lancelot Law. *Internal Factors in Evolution*. New York: George Braziller, 1965.

Williams, George C. *Adaptation and Natural Selection: A Critique of Some Current Evolutionary Thought*. Princeton: Princeton University Press, 1966.

Wills, Christopher. *The Wisdom of the Genes: New Pathways in Evolution*. New York: Basic Books, Inc., 1989.

Index

Abelev, G.I., 129
Adaptive pro-oncogenes, 35-41
  evidence for, 130-133
Agassiz, L. 186, 188
Aging, 77-78, 82-83, 100, 164
AIDS, 98, 140-142
Alexander, P., 90
Altman A.A., 133
Amateurs and science, 147-152, 186-187
Ames, B., 16, 29, 126, 183
Amphioxus, 131
Anders, F. 131
Andervont, H.B., 134
Andrew, W. 134
Anemones, 77
Angier, N., 30
Animal forbidden by theory, 168-170
Animal-manufacturing systems, 65-71
Animals
  avoidance of sunlight, 32, 86-92, 93-112
  number of cell types compared, 114
Annelids,
  regeneration in, 79
  compound eyes of, 94-95
Anti-oncogenes,
  33-35, 69, 90
  evidence for, 127-128
Appleman, P., 183
Arthropods [See insects]
Arthur, W., 10f, 71f
Asbestos, 144
Autectomy 26
Bacq, Z.M.. 90
Bank accounts compared to
  gene pools, 46-47
Bees, 45f
Bernhardt, P., 116.
Big bangs, biology's, 157, 190

Bilateria, 6f
Biologist's letter to author, 171, 174f
Bipedalism, 105-108
Birds, behavior, 73-74
Bishop, M. 183
Biston betularia, 20-21, 23, 49, 54-55, 69
Black boxes, 14-24
Bonner, J.T., 58, 78, 114f
Boveri, T., 126
Brain, human, origin, 105-108
Buffon, G.L., 151, 153
Butler, S., 72
Caenorhabditis,
  briggae, 82
  elegans, 75, 78
Cairns, J. 135, 140-143
Cambrian explosion, 73
Camouflage in butterfly, 34
Cancer,
  and xeroderma pigmentosum, 137
  as killer of children, 130
  definition, 15
  disease concept reconciled with
    theory, 175-176
  following trauma, 127, 144
  function of, identified, 27
  higher rates in more evolved animals, 132-133
  ignored by neo-Darwinism, 12
  in wild mice, 134
  in AIDS victims, 140-142
  in fish, 134
  increased after selection of adaptive genes, 35
  initiation by mutational event, 125-127
  initiation description, 18
  Kaposi's sarcoma, 141
  modern evidence supports theory,
  of human brain, 106

We owe our existence to cancer!

That is the startling assertion made by James Graham in his new theory of evolution.

Although the author's provocative synthesis was published in 1983 in the prestigious *Journal of Theoretical Biology*, professional evolutionists have ignored it. Concluding from their silence that they are incapable of evaluating his unorthodox theory, Graham--himself a self-taught amateur--decided to present it directly to the reading public. The result is this ingeniously persuasive book. In clear and nontechnical language, Graham argues for nothing less than the replacement of neo-Darwinism, the old theory of evolution, by his radical new theory.

The old theory claims that simple plant-evolution mechanisms were *completely* responsible for the evolution of human beings and all other complex animals. That, says Graham, is utter nonsense. Animal evolution *must* have involved something else, a mechanism that can account for the extraordinary precision used in the self-manufacture of animal bodies.

Graham then shows, coolly and methodically, that lethal cancer in juveniles--*cancer selection*--is the only plausible candidate for the missing mechanism. Despite the counter-intuitive nature of his astonishing proposition, within a very few pages the reader is compelled by the author's logic and the weight of the scientific evidence he musters (including the presence of cancer triggers in all normal cells) to agree that this most unsettling idea is not only plausible, but true.

Graham goes on to use his powerful theory to solve problems that confound